VOLUME TWO

ADVANCES IN STEM CELLS AND THEIR NICHES

Intestinal Stem Cell Niche

VOLUME TWO

ADVANCES IN STEM CELLS AND THEIR NICHES

Intestinal Stem Cell Niche

Edited by

DOMINIQUE BONNET
The Francis Crick Institute
London, United Kingdom

ELSEVIER

ACADEMIC PRESS
An imprint of Elsevier

Academic Press is an imprint of Elsevier
50 Hampshire Street, 5th Floor, Cambridge, MA 02139, United States
525 B Street, Suite 1800, San Diego, CA 92101-4495, United States
The Boulevard, Langford Lane, Kidlington, Oxford OX5 1GB, United Kingdom
125 London Wall, London EC2Y 5AS, United Kingdom

First edition 2018

ISBN: 978-0-12-813481-8
ISSN: 2468-5097

For information on all Academic Press publications
visit our website at https://www.elsevier.com/books-and-journals

Publisher: Zoe Kruze
Acquisition Editor: Zoe Kruze
Editorial Project Manager: Joanna Collett
Production Project Manager: James Selvam
Senior Cover Designer: Miles Hitchen

Typeset by SPi Global, India

CONTENTS

Contributors vii
Preface ix

1. Intestinal Stem Cells and Their Defining Niche **1**
Coralie Trentesaux and Béatrice Romagnolo

1. The Mammalian Intestinal Epithelium 1
2. ISC and Their Markers 3
3. Signaling Pathways Regulating Crypt Homeostasis 9
4. Organoid Culture and the Experimental Niche 17
5. The Epithelial Niche: Paneth Cells and Deep Crypt Secretory Cells 19
6. The Mesenchymal Niche 23
7. The Microbial Niche 25

References 27

2. Multitasking Paneth Cells in the Intestinal Stem Cell Niche **41**
Matthias Schewe and Riccardo Fodde

1. Paneth Cells: From Schwalbe and Paneth to the Discovery of Their Secretory Role 41
2. The Intestinal Crypt Stem Cell Niche 46
3. PCs, Dysbiosis, and IBD 60
4. Conclusive Remarks 67

References 68

3. Intestinal Stem Cells and Their Niche at Homeostasis and Under Stress **77**
Lucie Peduto, Philippe J. Sansonetti, and Giulia Nigro

1. The Gut: From Structure to Regeneration 77
2. Concept of Niche 83
3. Conclusions 92

References 93

4. Extracellular Matrix Remodeling in Intestinal Homeostasis and Disease **99**

Sang-Eun Lee, Isobel Massie, Laween Meran, and Vivian S.W. Li

1. Mammalian Intestine and the Surrounding Niche 100
2. ECM Dynamics in Intestinal Homeostasis and Regeneration 105
3. ECM Dynamics in IBD 111
4. ECM Dynamics in CRC 115
5. Tissue Engineering 122
6. Conclusion 127

Acknowledgments 128
References 128

CONTRIBUTORS

Riccardo Fodde
Erasmus MC Cancer Institute, Erasmus University Medical Center, Rotterdam, The Netherlands

Sang-Eun Lee
Stem Cell and Cancer Biology, The Francis Crick Institute, London, United Kingdom

Vivian S.W. Li
Stem Cell and Cancer Biology, The Francis Crick Institute, London, United Kingdom

Isobel Massie
Stem Cell and Cancer Biology, The Francis Crick Institute, London, United Kingdom

Laween Meran
Stem Cell and Cancer Biology, The Francis Crick Institute; Stem Cells and Regenerative Medicine Section, University College London Institute of Child Health, London, United Kingdom

Giulia Nigro
Unité de Pathogénie Microbienne Moléculaire, Institut Pasteur; INSERM U1202, Paris, France

Lucie Peduto
Unité Stroma, Inflammation & Tissue Repair, Institut Pasteur; INSERM U1224, Paris, France

Béatrice Romagnolo
INSERM U1016, Institut Cochin, CNRS UMR8104, Université Paris Descartes, Sorbonne Paris Cité, Paris, France

Philippe J. Sansonetti
Unité de Pathogénie Microbienne Moléculaire, Institut Pasteur; INSERM U1202; Chaire de Microbiologie et Maladies Infectieuses, Collège de France, Paris, France

Matthias Schewe
Erasmus MC Cancer Institute, Erasmus University Medical Center, Rotterdam, The Netherlands

Coralie Trentesaux
INSERM U1016, Institut Cochin, CNRS UMR8104, Université Paris Descartes, Sorbonne Paris Cité, Paris, France

PREFACE

This second volume of the "Advances in Stem Cells and Their Niches" series is dedicated to the *Intestinal Stem Cell Niche*.

The intestinal epithelium is a monolayer of cells covering the entire lumen of the gut that constitutes an important barrier against the external environment. Both small and large intestine are formed of crypt structure where intestinal stem cells (ISCs) reside.

The ISCs reside at the crypt base alongside the Paneth cells. Contrary to the majority of adult stem cells, ISCs divide actively and give rise to all different cell types, including enterocytes, goblet cells, enteroendocrine cells, tuft cells, as well as Paneth cells.

ISCs in the crypt base are maintained by their surrounding niche for precise regulation of self-renewal and differentiation. The intestinal stem cell niche is composed of Paneth cells, stromal cells, smooth muscle cells, and neural cells as well as the extracellular matrix components.

In the first chapter, Drs. Trentesaux and Romagnolo provide a short historical summary of the progress made in our understanding and characterization of the ISCs, the regulatory niche signals, and the different cell types contributing to this niche in the adult intestine, touching briefly on ISC organoid culture.

In the second chapter, Drs. Schewe and Fodde highlight the multifunctional role played by the Paneth cells in the regulation of the ISCs. They describe how Paneth cells via sensing environmental factors like dietary changes and/or inflammatory stimuli are capable of translating these different signals into instructive pathways regulating ISCs activities.

In the third chapter, Dr. Peduto et al. describe the role of the different cell types constituting the ISC niche and more specifically discuss the recent evidence, demonstrating the involvement of gut microbiota in the regulation of ISCs fate during homeostasis and under stress conditions. In the last chapter, Dr. Lee et al. shed light into the role played by the extracellular matrix components in the regulation of ISCs in both homeostasis and during pathogenesis including inflammatory bowel disease or cancer. They also discussed some of the new development in tissue engineering using biological scaffold for regeneration medicine and disease modeling.

This book is thus not only providing an excellent description of the cellular and extracellular matrix components constituting the ISC niche, it also described the signaling pathways involved in the ISCs regulation during homeostasis, under stress, and during disease progression. Lastly, this book highlights the new role of the microbiota in the ISC regulation and the growing multifunction of Paneth cells.

This book should thus be of interest for academic researchers, postdoctoral research fellows, and graduate students working on adult stem cells, cell signaling mechanisms, tissue engineering, and cancer development.

Dominique Bonnet
The Francis Crick Institute, London, United Kingdom

CHAPTER ONE

Intestinal Stem Cells and Their Defining Niche

Coralie Trentesaux, Béatrice Romagnolo[1]
INSERM U1016, Institut Cochin, CNRS UMR8104, Université Paris Descartes, Sorbonne Paris Cité, Paris, France
[1]Corresponding author: e-mail address: beatrice.romagnolo@inserm.fr

Contents

1. The Mammalian Intestinal Epithelium 1
2. ISC and Their Markers 3
 2.1 Historical Models of ISC 3
 2.2 Markers for CBC Cells 5
 2.3 Markers for +4 Stem Cells 6
 2.4 Current View of ISC 7
 2.5 Progenitors as Potential ISC 8
3. Signaling Pathways Regulating Crypt Homeostasis 9
 3.1 Wnt/β-Catenin Signaling 9
 3.2 Hedgehog, BMP, and PI3K/PTEN Signaling 14
 3.3 Notch Signaling 15
4. Organoid Culture and the Experimental Niche 17
5. The Epithelial Niche: Paneth Cells and Deep Crypt Secretory Cells 19
6. The Mesenchymal Niche 23
7. The Microbial Niche 25
References 27

1. THE MAMMALIAN INTESTINAL EPITHELIUM

The adult mammalian intestinal epithelium consists of a single layer of cells folded upon itself creating cavities called crypts of Lieberkühn surrounded by connective mesenchyme. In the small intestine, the crypts are almost entirely composed of proliferative cells, with the exception of Paneth cells, a lineage first recognized for their ability to secrete antimicrobial peptides (AMP) that protect the crypt environment. As will be discussed later, these cells are also largely involved in the intestinal stem cell (ISC) niche.

Advances in Stem Cells and their Niches, Volume 2
ISSN 2468-5097
https://doi.org/10.1016/bs.asn.2017.12.003

Larger epithelial folds called villi protrude outward into the intestinal lumen and contain exclusively differentiated cells. These are primarily enterocytes, which make up over 80% of intestinal epithelial cells and whose function is to absorb dietary nutrients from the lumen. In addition, villi contain goblet cells that secrete a protective layer of mucus around the epithelium and enteroendocrine cells that produce hormones (Gordon, 1989). Three rarer and less-characterized cell types are also found within the differentiated epithelium: tuft cells (Gerbe et al., 2011), cup cells (Madara, 1982), and microfold (M) cells associated with lymphoid follicles called Peyer's patches (Kraehenbuhl & Neutra, 2000). Along the cephalocaudal axis of the small intestine, from the duodenum to the ileum, the length of the villi diminishes and the amount of secretory cells namely goblet cells and Paneth cells, increases (Cheng, 1974; Darmoul & Ouellette, 1996). This gradient accompanies a diminishing need for absorption and a considerable increase in microbe density (Savage, 1977). The colon, where stool is compacted for excretion, has a flat surface with no villi; the bottom third of colonic crypts contains proliferative cells while the upper crypt is differentiated. Microbe density is at its greatest in this segment of the intestine and, accordingly, the proportion of mucus-secreting goblet cells is considerably higher (Savage, 1977). Paneth cells, however, are notably absent in the colon.

The adult mouse intestine contains approximately 1.1 million crypts, with around two-thirds of the cells of each crypt dividing every 12h, producing over 300 million new cells each day (Gordon, Schmidt, & Roth, 1992). Three to 16 (depending on the study) actively dividing ISC, called crypt basal columnar cells (CBC) reside at the crypt bottom alongside Paneth cells in the small intestine and related secretory cells in the colon (Altmann, 1983; Bjerknes & Cheng, 1981b; Potten & Loeffler, 1987; Sasaki et al., 2016; Sato, van Es, et al., 2011). CBC give rise both to new CBC and to progenitors called transit amplifying (TA) cells that will rapidly proliferate and progressively differentiate as they migrate up the crypt. TA daughters will either be absorptive lineage progenitors, giving rise to enterocytes, or secretory progenitors that will give rise to goblet, enteroendocrine, and Paneth cells (Bjerknes & Cheng, 1999; Schmidt, Wilkinson, & Ponder, 1985). In the small intestine, cells become terminally differentiated upon reaching the villus, then continue their columnar migration to the tip of the villus, where they are expelled and die in a process termed anoikis. Paneth cells, unlike the other progenitor lineages, terminally differentiate as they migrate down to the bottom of the crypt. The position of a Paneth cell is therefore linked to its maturity, with the oldest and more mature

Paneth cells found at the crypt base (Bjerknes & Cheng, 1981a). The entire process of proliferation, differentiation, and expulsion of intestinal epithelial cells takes only 4–5 days (Cheng & Leblond, 1974b; Wright & Irwin, 1982). The 6–8-week lifespan of Paneth cells further distinguishes them from other differentiated cell types of the intestinal epithelium (Ireland, Houghton, Howard, & Winton, 2005).

Epithelial patterning occurs not just along the crypt-villus axis, but also along the cephalocaudal axis. Fetal small intestinal isografts implanted subcutaneously develop into a histologically normal epithelium with a well-developed lamina propria, enteric nervous system, and smooth muscle layer. Strikingly, despite the absence of luminal contents, secretory goblet cells and Paneth cells are present, and the grafted tissue undergoes active peristalsis. Furthermore, the proportion and distribution of epithelial cell types as well as the gene expression profile of the isografts match those of the normal tissue segment it came from (Rubin, Roth, Birkenmeier, & Gordon, 1991; Rubin, Swietlicki, Roth, & Gordon, 1992). This suggests that ISC somehow retain a memory of their position-specific program. Region-specific gene programs are also conserved in cultured organoids from murine or human crypts (Middendorp et al., 2014), indicating that this "memory" is intrinsic to the epithelium. Strengthening this notion, small intestinal organoids reimplanted into colonic epithelium reconstitute a self-renewing epithelium with features of the small intestine (Fukuda, Mizutani, & Mochizuki, 2014).

2. ISC AND THEIR MARKERS

2.1 Historical Models of ISC

The identity of ISC has remained a subject of debate over the past decades. Two historical models of ISC have been originally proposed but direct evidence to illustrate their stemness was put forward only in the last decade. Pioneer work by Cheng, Bjerknes, and Leblond using electron microscopy, ^{3}H-thymidine incorporation, and clonal mutagenesis demonstrated the presence of CBC, undifferentiated proliferative cells capable of giving rise to the different intestinal epithelial lineages (and therefore said to be multipotent), at positions 1–4 from the bottom of the intestinal crypt (Bjerknes & Cheng, 1999, 2002; Cheng & Leblond, 1974a). At the same time, works led by Potten recognized cells at the fourth position from the crypt bottom (+4) as capable of long-term ^{3}H-thymidine label retention

(Potten, Hume, Reid, & Cairns, 1978), an established property of tissue stem cells. Without specific molecular markers to isolate them, it remained unclear whether the mitotically active CBC or the relatively quiescent label-retaining cells (LRC) ought to be considered as the bona fide ISC (Fig. 1).

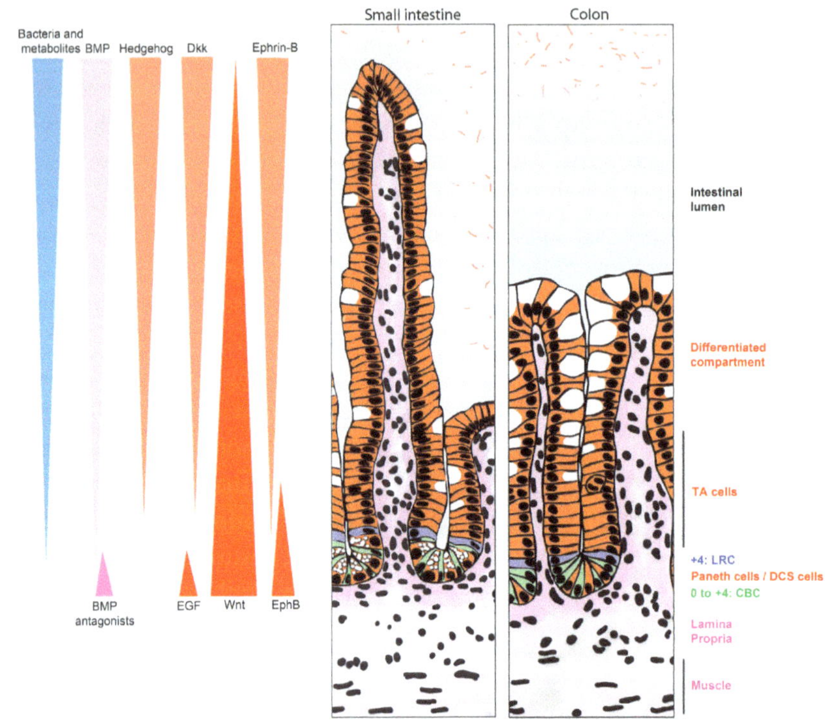

Fig. 1 Morphology of the small intestinal and colonic epithelium and niche signals involved in the regulation intestinal stem cells (ISC). The intestinal epithelium consists of a single layer of cells divided into two compartments: the differentiated compartment (villi in the small intestine, crypt tops in the colon) primarily contain absorptive enterocytes and secretory goblet cells while the proliferative crypts harbor transit amplifying (TA) cells that progressively differentiate as they migrate up the crypt, secretory niche cells (Paneth cells in the small intestine, deep crypt secretory (DCS) cells in the colon), and ISC. Two types of ISC have been proposed: crypt basal columnar cells (CBC) reside in positions 0 to +4 from the crypt bottom, intercalated between Paneth or DCS cells in the small intestine and colon, respectively. CBC are actively cycling and classically recognized by their marker Lgr5. More quiescent, label-retaining cells (LRC) at the +4 position expressing markers like Bmi1 have been proposed as reserve ISC. Several signals—most notably canonical Wnt signaling—regulate ISC proliferation, differentiation, and response to microenvironmental changes. These signals come from epithelial cells, particularly Paneth or DCS cells, from mesenchymal cells and extracellular matrix components present in the lamina propria, as well as from luminal nutrients, bacteria, and their metabolites. All these components and signals constitute the ISC niche.

2.2 Markers for CBC Cells

The Clevers lab and collaborators identified Lgr5 as the first marker for CBC. *Lgr5* encodes a leucine-rich repeat-containing G protein-coupled transmembrane receptor first identified as a Wnt/β-catenin target gene. Although Lgr5 labeling has proven difficult due to its low expression levels, the development of a mouse model in which an EGFP and tamoxifen-inducible Cre-recombinase cassette was knocked in at the *Lgr5* locus (Lgr5–EGFP–CreER^{T2}) allowed characterization of the Lgr5^{+} cells. In the adult small intestine, Lgr5^{+} cells are located at the base of the crypt, intermingled with Paneth cells. They are rapidly cycling since within 24 h almost all Lgr5^{+} cells of a crypt undergo mitosis. In the colon, Lgr5^{+} cells are also found at the crypt bottom interspersed with secretory cells but have slower cycling kinetics than in the small intestine. The generation of in vivo lineage tracing from Lgr5^{+} cells using an Lgr5–EGFP–CreER^{T2}/Rosa26lacZ mouse model indicated that Lgr5^{+} cells transmit lacZ staining to their progeny and repopulate the entire epithelium during homeostasis. Lgr5^{+} cells are long lived and contribute to tissue renewal over the entire lifetime of the mouse. Therefore, Lrg5 is a reliable marker of CBC (Barker et al., 2007), and Lgr5–EGFP–CreER^{T2} mice a valuable tool to understand their regulation. The ultimate proof of Lgr5^{+} CBC as ISC is the capacity of single isolated Lgr5^{+} cells cultured in vitro to generate structured intestinal organoids that contain the four major differentiated lineages of the intestinal epithelium and that self-renew, as shown by continuous passages (Sato et al., 2009). A CBC stem cell signature using gene expression and proteome profiling has been determined using FACS-purified Lgr5^{+} cells from the Lgr5–EGFP–CreER^{T2} model. A comparison between Lgr5^{+} CBC and their progeny sorted by different levels of EGFP (high in CBC and low in their progeny) provided a comprehensive Lgr5^{+} ISC signature of approximately 500 genes (Muñoz et al., 2012). The signature contains many Wnt/β-catenin modulators and target genes such as *Lgr5*, *Sox9*, *Ascl2* (achaeate-scute homolog 2), *EphB2*, *Troy/Tnfrsf19*, *Axin2*, *Znrf43* (zinc and ring finger 43) (Hao et al., 2012), and *Rnf3* (ring finger 3) (Koo et al., 2012). This and subsequent Lgr5^{+} cell signatures (Muñoz et al., 2012; van der Flier, van Gijn, et al., 2009) along with candidate-approach studies also provided several other verified CBC markers. Among the new markers identified, some have been confirmed as ISC markers by lineage tracing: *Smoc2* (SPARC-related modular calcium binding 2), encoding a BMP (bone morphogenic protein) signaling inhibitor (Muñoz et al., 2012), *Musashi-1*, encoding an RNA-binding protein (Potten et al., 2003), *Prominin-1* (Zhu et al., 2009) and notably *Olfm4*, which is highly expressed compared to other mentioned CBC markers and allows easy detection of CBC (Schuijers, Van Der Flier, Van Es, & Clevers, 2014; van der Flier, Haegebarth, Stange, van de Wetering, & Clevers, 2009). Ascl2, a basic helix–loop–helix transcription factor, was identified as a master regulator of CBC maintenance. The conditional deletion of *Ascl2* results in the loss of CBC and its overexpression promotes drastic expansion of the ISC compartment (Reed et al., 2012; Schuijers et al., 2015; van der Flier, van Gijn, et al., 2009).

The identification of these CBC markers has allowed better characterization of ISC during homeostasis but also in disorders like colorectal cancer (CRC). In light of the high number of proliferative cells in the intestine and comparatively low occurrence of spontaneous intestinal tumors, it has long been proposed that only mutagenic hits accumulated in ISC could result in tumor development. Although it has proven difficult to ascertain which cell types bear the initial tumor-initiating mutations in humans, recent studies in mice suggest that CBC are likely to be the CRC cell of origin (Barker et al., 2009). Direct evidence has been obtained by the aberrant activation of Wnt signaling upon the inactivation of tumor suppressor Apc in $Lgr5^+$ CBC, leading to a rapid onset of adenomas. In contrast, deletion of *Apc* in more differentiated cells results only in ectopic proliferative foci which rarely progress to microadenomas. In addition, lineage tracing of $Lgr5^+$ adenoma cells revealed that they fuel the growth of adenoma by generating all the cell types present in the adenoma indicating that they act as multipotent stem cells of the adenoma (Schepers et al., 2012). Many CBC markers including *Lgr5*, *Musashi-1*, *Ascl2*, *Olfm4*, and *EphB2* are expressed both in human CRC and in genetically induced murine adenomas (Fan et al., 2010; Jang et al., 2016; Jubb et al., 2006; Li et al., 2015; Potten et al., 2003).

2.3 Markers for +4 Stem Cells

The first +4 stem cell (+4SC) marker investigated by lineage tracing was Bmi1, a polycomb-repressing complex member essential in hematopoietic and neural stem cell self-renewal. In intestinal homeostasis, *Bmi1* is expressed at the +4 position in a minority of small intestinal crypts (around 10%) and particularly in the duodenum. By in vivo lineage tracing using a Bmi1-$CreER^{T2}$/Rosa26lacZ mouse model and organoid culture experiments with sorted cells, $Bmi1^+$ cells were shown to self-renew and give rise to the different epithelial lineages but with much slower cycling kinetics compared to $Lrg5^+$ CBC (Sangiorgi & Capecchi, 2008; Yan, Chia, & Li, 2012).

In addition, ablation of Bmi1$^+$ cells using targeted expression of diphtheria toxin (DT) caused crypt loss consistent with stem cell defect.

Ablation of the Lgr5$^+$ CBC using targeted expression of the DT receptor (DTR) caused an increase of Bmi1$^+$ cells which can regenerate the Lgr5$^+$ stem cell population and maintain intestinal homeostasis (Tian et al., 2011). Thus, these relatively quiescent +4SC, reminiscent of the LRC described by Potten, have been considered as reserve stem cells in case of damage to the active CBC stem cells. In support of the mobilization of the +4SC following injury, such as high dose of radiation (up to 12 Gy), Lgr5$^+$ cells are rapidly lost whereas Bmi1$^+$ cells expand and allow epithelial recovery though replenishment of Lgr5$^+$ cells (Yan et al., 2012). In fact, Bmi1 has been shown to be involved in DNA damage response signaling; it may therefore play a direct role in the resistance of reserve stem cells to irradiation (Ginjala et al., 2011; Ismail, Andrin, McDonald, & Hendzel, 2010; Pan, Peng, Hungs, & Lin, 2011).

However, Bmi1-expressing cells are only present in a small proportion of crypts in the proximal small intestine and are not responsible for the post-injury regeneration throughout the distal small intestine or colon in these contexts (Tian et al., 2011; Yan et al., 2012). Other markers must therefore be used to identify the reserve stem cells in these regions. Coincidentally, *Tert* (telomerase reverse transcriptase) also labels single cells at the +4 position that are slowly cycling, label retaining, resistant to radiation, and capable of regenerating the epithelium and Lgr5$^+$ cells following injury (Breault et al., 2008; Montgomery et al., 2011). Other reported markers of +4SC include *Hopx* (homeodomain only) (Takeda et al., 2011), *Lrig1* (leucine-rich repeats and immunoglobulin-like domains 1) (Powell et al., 2012), and, debatably, *Dcamkl1* (doublecortin-like kinase 1) (May et al., 2008, 2009). Remarkably, cells expressing these markers can give rise to CBC and, conversely, Lgr5$^+$ cells can give rise to +4SC (Takeda et al., 2011).

2.4 Current View of ISC

The mutually exclusive nature of these two proposed stem cell populations, CBC and +4SC, has been extensively challenged in the literature in the past few years and appears less straightforward than initially thought. Single molecule in situ hybridization illustrates the restricted expression of certain markers like *Lgr5*, *Msi1*, *Ascl2*, *Olfm4*, and *EphB3* at the crypt bottom, but also that markers like *Prominin-1* and *Lrig1* have a broader pattern of expression along the crypt axis including both the majority of Lgr5$^+$ and

TA cells (Itzkovitz et al., 2012). Likewise, Muñoz et al. clearly demonstrate that +4SC markers *Bmi1*, *Tert*, and *Hopx* are enriched in $Lgr5^+$ CBC. This has been supported by several approaches both at the mRNA (Grün et al., 2015; Muñoz et al., 2012; van der Flier, van Gijn, et al., 2009) and protein levels (Schepers, Vries, van den Born, van de Wetering, & Clevers, 2011; Wong et al., 2012), indicating that these markers may not be specific to quiescent stem cells and that the exact relationship between CBC and +4SC still needs to be further defined.

2.5 Progenitors as Potential ISC

The plasticity of progenitor cells represents another source of potential ISC. One study demonstrated that secretory progenitors expressing Delta-like (Dll) 1, a Notch ligand, can revert to multipotent stem cells. Using a Dll1–GFP–$CreER^{T2}$/Rosa26lacZ mouse model, lineage tracing indicated that secretory progenitors can revert to $Lgr5^+$ ISC upon ablation of CBC by irradiation and form $Lgr5^+$ ISC-containing organoids ex vivo in the presence of Wnt signals (van Es, Haegebarth, et al., 2012; van Es, Sato, et al., 2012). Similarly, Buczacki and colleagues used and elegant model to mark only LRC by using a fluorescent-labeled histone H2B reporter construct (H2B-YFP) (Buczacki et al., 2013). They demonstrated that during homeostasis LRC express +4SC markers including *Lrig1*, *Tert*, and *Hopx*, but also *Lgr5* and markers of early Paneth cells and enteroendocrine cells. Using an ingenious lineage tracing approach, they demonstrated that during homeostasis LRC remain quiescent and express markers of Paneth cells. However, upon epithelial injury induced by irradiation or chemotherapy treatment, LRC proliferate, and act as ISC to give rise to all differentiated cell lineages. This study therefore indicates that committed secretory precursors are a subpopulation of $Lgr5^+$ cells which can act as a potential ISC pool upon injury. More recently, enterocyte progenitors have been shown to provide an alternative source of potential ISC. Using lineage tracing based on intestinal alkaline phosphatase (Alpi), a marker for progenitors and differentiated enterocytes, Tetteh et al. demonstrated that upon eradication of $Lgr5^+$ cells through DT treatment, the proliferating crypt $Alpi^+$ cells and not differentiated villus $Alpi^+$ cells were able to dedifferentiate and to produce all differentiated cell types, thereby acting as an alternative source of ISC (Tetteh et al., 2016). Two new studies from the groups of Kuo and Shivdasani categorize *Bmi1*-expressing cells as enteroendocrine progenitors, and further demonstrate the capacity of progenitors from this lineage to dedifferentiate and replenish the $Lgr5^+$ ISC pool in case of injury (Jadhav et al., 2017; Yan et al., 2017). Taken together, all these studies call into question the existence of a dedicated reserve stem cell population and rather highlight the plasticity of the intestinal epithelium along with the importance of maintaining the active ISC pool. Questions regarding the molecular drivers and the degree of this plasticity remain open. As a hypothesis, it could be argued that upon injury leading to loss of CBC, committed progenitors regain direct contact with the stem cell niche, which may participate to their dedifferentiation and revert them to stemness.

More than a way to visualize ISC and analyze their contributions to intestinal pathology, these markers allow for a better understanding of the interaction of ISC with their surroundings. Many questions regarding the functional relation between ISC and their direct microenvironment, or niche, can now be approached. What molecular signals are involved in crypt homeostasis and how do they affect ISC? What cell types constitute the niche and provide these signals? How are ISC capable of sensing and responding to microenvironmental changes? What delimits the "positional address" of ISC? Are ISC actively anchored to this position or, on the contrary, does the niche define "stemness"?

3. SIGNALING PATHWAYS REGULATING CRYPT HOMEOSTASIS

Proliferation, differentiation, and migration along the crypt-villus axis are tightly regulated by an array of interconnected signaling pathways. Both epithelial signals and epithelial–mesenchymal crosstalk act in concert to create a complex molecular environment supporting ISC activity. Below, we discuss the principal and best-characterized pathways essential to ISC homeostasis and the pathological consequences of their deregulation.

3.1 Wnt/β-Catenin Signaling

The major regulatory pathway of the ISC compartment is the Wnt/β-catenin signaling pathway. In the absence of Wnt signals, free cytosolic β-catenin is recognized by a protein complex containing the tumor suppressors Apc (adenomatous polyposis coli), Axin1, CK1α (casein kinase 1α), and GSK-3β (glycogen synthase 3β), called the β-catenin destruction complex. Indeed, CK1α and GSK-3β will phosphorylate serine and threonine residues of the β-catenin N-terminus, leading to its ubiquitination by E3 ubiquitin ligase β-TrCP and subsequent proteasomal degradation. Wnt ligands bind a

Frizzled (Fzd) receptor and coreceptor lipoprotein-related protein (Lrp5/6 in the intestine). This interaction triggers a cascade of events involving phosphorylation of disheveled and membrane sequestration of Axin1 which, through mechanisms that remain unclear, inhibits the destruction complex, and prevents β-catenin degradation. β-Catenin is therefore free to accumulate and translocate to the nucleus, where it will bind transcription factor Tcf/Lef (T-cell factor/lymphoid enhancer factor) to induce Wnt/β-catenin target genes (Logan & Nusse, 2004). Numerous studies have illustrated that CBC homeostasis is dependent on the activity of the Wnt/β-catenin signaling cascade. Wnt/β-catenin signal strength is dependent on several agonists and antagonists which may participate to the very context-dependent transcriptional response of the pathway. In CBC, Lgr5 and its homologs, Lgr4 and Lgr6, constitute the receptors for R-spondins, potent Wnt signal enhancers and CBC growth factors that work by neutralizing Rnf43 and Znrf3, two transmembrane E3 ligases that remove Wnt receptors from the cell surface (de Lau et al., 2011).

β-Catenin also plays a role at cell adhesion complexes, linking E-cadherin to α-catenin and the actin cytoskeleton. Consequently, throughout the intestinal epithelium, β-catenin is mainly localized to the plasma membrane between epithelial cells, but also displays strong nuclear staining at the crypt bottom. Indeed, Wnt/β-catenin signaling is particularly active in CBC and Paneth cells and exhibits a decreasing gradient of activity from the crypt bottom up. The major indicator of the role of Wnt/β-catenin signaling in the intestinal epithelium is the extensive effects of its deregulation. Patients with familial adenomatous polyposis (FAP) carry a germline mutation in one *APC* allele. They develop hundreds to thousands polyps in the large intestine and are at high risk for developing CRC (Miyoshi et al., 1992). Genetically engineered mouse models have been useful for modeling FAP; the adenomas that arise in *Apc*-mutant mice are similar to those in FAP patients (Taketo & Edelmann, 2009). *Apc*-mutant mice develop large numbers of adenomas in their small intestine and fewer in the large intestine which rarely progress to invasive adenocarcinoma, likely linked to the short lifespan of these mice (Colnot et al., 2004). Use of a conditional *Apc* allele showed that loss of Apc is a critical event to initiate intestinal tumor development. Conditional deletion of both alleles of *Apc* specifically in the intestinal epithelium leads to β-catenin relocalization to the nucleus, intense cell proliferation in the crypt compartment, along with impaired migration and differentiation throughout the epithelium (Andreu et al., 2005; Sansom et al., 2004). Moreover, a vast majority of human sporadic CRC carry early driver mutations of *APC* or, less frequently, mutations on other genes encoding components of the canonical Wnt pathways like Axin2, R-spondins, Rnf43, or β-catenin itself (Fodde et al., 2001; Koo et al., 2012; Morin et al., 1997; Powell et al., 1992; Seshagiri et al., 2012). These tumors have a high level of nuclear β-catenin, and the intestine-specific expression of β-catenin mutants lacking its N-terminal phosphorylation sites in mice also leads to the development of such lesions (Harada et al., 1999; Romagnolo et al., 1999). The protumorigenic effects of *Apc* mutation therefore result in the constitutive activation of canonical Wnt/β-catenin signaling, which drives tumorigenesis. The fact that Apc deficiency acts as a driver and not just an initiating event in colorectal tumorigenesis has recently been underlined by Dow et al. In their study, restoring *Apc* in established tumors, even those carrying additional *Kras* and *Tp53* mutations, leads to their regression and the reestablishment of normal epithelial architecture (Dow et al., 2015).

Multiple studies have demonstrated that the Wnt/β-catenin pathway is essential for the homeostatic maintenance and proliferation of CBC. In the intestine, disruption of the pathway by deletion of either *Tcf4* or the β-catenin gene *Ctnnb1* results in a fatal disruption of intestinal architecture with ablation of the proliferative crypts (Fevr, Robine, Louvard, & Huelsken, 2007; Ireland et al., 2004; Korinek et al., 1998; van Es, Haegebarth, et al., 2012; van Es, Sato, et al., 2012). More upstream disruption of the pathway by ectopic expression of Wnt antagonist Dickopff (Dkk) 1 also results in the loss of stem and progenitor cells (Kuhnert et al., 2004; Muncan et al., 2006; Pinto, Gregorieff, Begthel, & Clevers, 2003). Conversely, overstimulation of the pathway by injecting recombinant human R-spondin1 into mice induces rapid crypt proliferation involving β-catenin stabilization (Kim et al., 2005). Accordingly, conditional deletion of both *Znrf3* and *Rnf4* expands proliferation in the crypt and is protumorigenic (Hao et al., 2012; Koo et al., 2012). Although deletion of *Lgr5* has no effect on intestinal cell proliferation, inactivation of *Lgr4* severely decreases crypt cell proliferation. Genetic deletion of *Lgr4* and *Lgr5* together completely abolishes proliferation and confirms that R-spondins act as major drivers of crypt self-renewal (de Lau et al., 2011). $Lgr5^{+}$ CBC were shown to be highly responsive to such modulations in Wnt signaling, expanding upon stimulation by R-spondin1 or Apc loss and dying upon Dkk1-mediated inhibition of the pathway (Barker et al., 2009; Yan et al., 2012). A recent report demonstrated that Ascl2, a Wnt target gene expressed in CBC, acts as a master regulator of the Lgr5 ISC transcriptional program: Ascl2 cooperates with β-catenin/Tcf4 complexes to switch on transcription of signature genes that define CBC identity (Schuijers et al., 2015).

In addition to proliferation, Wnt/β-catenin signaling also affects differentiation of ISC daughters. *Apc*-deficient murine tumors overexpress Paneth cell-specific genes, and this surprisingly also applies to human colorectal tumors despite the absence of these cells in the healthy colonic mucosa (Andreu et al., 2008; Crawford et al., 1999; van Es, Jay, et al., 2005; van Es, van Gijn, et al., 2005). Furthermore, in this context, Paneth cells are no longer restricted to the crypt bottom and while they do express the AMP cryptdin, they do not present mature secretion granules, suggesting that their differentiation is incomplete. The fact that Paneth cells express a unique Wnt receptor, Fzd5, as opposed to other epithelial cells, which express Fzd6, supports a distinct Wnt response in these cells (van Es, Jay, et al., 2005; van Es, van Gijn, et al., 2005). Intestine-specific deletion of *Fzd5* also results in mislocalized Paneth cells lacking nuclear β-catenin as well as *Cryptdin* and *EphB3* expression. In fact, Paneth cell-restricted expression of genes encoding Cryptdins, EphB3, and matrix metalloprotease (Mmp) 7 is transcriptionally regulated by β-catenin and Tcf4 (Andreu et al., 2008; Crawford et al., 1999; van Es, Jay, et al., 2005; van Es, van Gijn, et al., 2005). Precise control of Wnt signaling, in addition to its importance in proliferation and ISC cell maintenance, is therefore also essential for Paneth cell differentiation and maturation (Andreu et al., 2008; van Es, Jay, et al., 2005; van Es, van Gijn, et al., 2005).

The scattered localization of Paneth cells upon deregulation of Wnt signaling hints at its implication in migration along the crypt-villus axis as well. *EphB3*- or *Ephrin-B1*-knockout intestines also display delocalized Paneth cells (Batlle et al., 2002; Cortina et al., 2007). Ephrin-B ligands and their EphB receptors are cell sorting molecules notably involved in axon guidance in the nervous system. In accord with their transcriptional activation by β-catenin, expression of EphB2 and EphB3 is the highest at the crypt bottom. Specifically, Paneth cells express EphB3, $Lgr5^+$ cells express both EphB3 and EphB2, and TA cells express decreasing levels of EphB2 as they migrate up the crypt. Expression of the Ephrin-B ligands is repressed by β-catenin, so it follows the opposite pattern with the highest expression in differentiated cells and the lowest expression at the crypt bottom. Therefore, as cells migrate up the crypt and Wnt signaling decreases, they gradually decrease *EphB2* and increase *Ephrin-B* expression. Since the interaction between EphB receptors and Ephrin-B ligands has a repellant effect, this mechanism keeps EphB-expressing Paneth and ISC restricted to the crypt bottom and prevents the downward migration of differentiating progenitors (Batlle et al., 2002). Knockout of both *EphB2* and *EphB3* in mice therefore results in dispersion of proliferative cells throughout the epithelium, intermingled with the differentiated cells. The effects of EphB/Ephrin-B interaction are thought to act, at least in part, through the regulation of tight junctions via activation of ADAM (a disintegrin and metalloproteinase domain) family mediated cleavage of E-cadherin at interfaces between neighboring EphB- and Ephrin-B-expressing cells (Solanas, Cortina, Sevillano, & Batlle, 2011). Despite the overactivation of Wnt signaling in CRC, *EphB* expression is often silenced at the adenoma–carcinoma transition, promoting cancer progression through a mechanism dependent on E-cadherin-mediated adhesion (Batlle et al., 2005; Cortina et al., 2007).

β-Catenin can also directly affect cell adhesion and thereby migration through its direct interaction with E-cadherin. In fact, β-catenin was historically first recognized for its role in adherens junction complexes. Modulation of E-cadherin levels affects migration speed and organization along the crypt-villus axis—with overexpression of the E-cadherin gene *Cdh1* inducing slower, structured migration and loss of E-cadherin resulting in disordered, rapid migration (Bondow, Faber, Wojta, Walker, & Battle, 2012; Hermiston, Wong, & Gordon, 1996; Schneider et al., 2010). Similarly, overexpression of *Apc* results in disordered migration (Wong, Hermiston, Syder, Gordon, & Kipnis, 1996). Proliferation in the crypt is also inversely linked to E-cadherin levels (Hermiston et al., 1996) and β-catenin is known to interact with E-cadherin and Tcfs through an overlapping binding region, suggesting that the two functions could be in stoichiometric competition (Choi, Huber, & Weis, 2006; Graham, Weaver, Mao, Kimelman, & Xu, 2000; Huber & Weis, 2001). Supporting this notion, high expression of E-cadherin correlates with lower tumorigenic potential of mutant β-catenin (Huels et al., 2015). It has even been proposed in cell culture models that E-cadherin complexes could modulate more upstream Wnt receptor signaling (Maher, Flozak, Stocker, Chenn, & Gottardi, 2009). Contrarily, intercellular junctions appear unaffected by the deletion of β-catenin in the intestinal epithelium (Fevr et al., 2007). Evidently, the link between the opposing functions of β-catenin in Wnt signaling-mediated proliferation and E-cadherin-dependent cell adhesion still needs to be untangled.

The Wnt activity gradient from the crypt bottom is in part established by the localized production of Wnt ligands. Crypt epithelial cells produce Wnt3, Wnt9b, and Wnt6 (Farin, Van Es, & Clevers, 2012; Gregorieff et al., 2005). Of these, Wnt3 is specifically expressed in Paneth cells. The intestinal mesenchyme also produces Wnt ligands, namely Wnt2b, Wnt4, and Wnt5a, which have been linked to noncanonical Wnt signaling. Crypt epithelial cells in the small intestine and colon also express various Fzd receptors as well as different Tcfs. Furthermore, the mesenchyme near the differentiated epithelium also produces Wnt antagonists like Dkk in a localized manner (Gregorieff et al., 2005). All this combined makes for a rather intricate and complex regulation of Wnt signaling resulting in a Wnt activity-permissive niche in the stem cell and Paneth cell compartment and repression of canonical Wnt signaling in the differentiated epithelium.

3.2 Hedgehog, BMP, and PI3K/PTEN Signaling

Hedgehog-family proteins and BMPs link epithelial homeostasis to mesenchymal signals. The Hedgehog receptor, Patched (Ptch), normally inhibits the constitutively active smoothened (Smo) receptor, thereby permitting cleavage of Gli transcription factors (Gli1, 2, and 3) into their repressive forms. Upon Hedgehog binding, Ptch is repressed, and Smo signaling allows Gli-mediated transcription of Hedgehog target genes. Hedgehog signaling is exclusively paracrine in the adult intestine; Indian Hedgehog and Sonic Hedgehog (Ihh and Shh, the main Hedgehog ligands in the intestine) secreted by enterocytes binds Ptch receptors on neighboring mesenchymal cells, inducing BMP production from these cells (Madison et al., 2005; van Dop et al., 2009). Binding of BMP to its type 2 receptor, Bmpr2, results in the phosphorylation and activation of its type 1 receptor, Bmpr1, which subsequently phosphorylates Smad proteins 1, 5, or 8. These will then heterodimerize with Smad4 before translocating to the nucleus to induce transcription of target genes. In the intestine, BMP2 and BMP4 are produced by intravillus and intercryptic mesenchymal cells, and Bmpr1 is expressed in differentiated cells and ISC but not in the proliferative TA cells (Haramis et al., 2004; He et al., 2004). BMP antagonists like Noggin, Gremlin1, Gremlin2, or Chordin-like 1 are expressed in subcryptic myofibroblasts and smooth muscle (He et al., 2004; Kosinski et al., 2007). It thus follows that BMP activity affects only in villus epithelial cells. The constitutive activation of Hedgehog signaling results in increased BMP signaling and depletion of proliferative and stem cells (van Dop et al., 2009). Conversely, either inhibition of Hedgehog or deregulation of the BMP pathway results in expansion of the ISC compartment, excessive crypt formation, and the formation of hamartomatous polyps (Davis et al., 2015; Haramis et al., 2004; He et al., 2004; Madison et al., 2005; Qi et al., 2017). These lesions are typical of those found in patients with juvenile polyposis syndrome, often associated with *BMPR1A* or *SMAD4* mutations. Therefore, Hedgehog and BMP signaling restrain proliferation and promote differentiation, in opposition with Wnt/β-catenin signaling.

In fact, modulation of either Hedgehog or BMP signaling affects nuclear β-catenin levels and the expression of its target genes (Madison et al., 2005; van den Brink et al., 2004; van Dop et al., 2009). Explaining this phenomenon, BMP signaling was shown to directly interfere with Wnt-mediated β-catenin activation through a mechanism involving PI3K and PTEN activation and subsequent inhibition of Akt (He et al., 2004; Waite, Waite, & Eng, 2003). Similar to disruption of BMP signaling, deletion of PTEN leads to de novo crypt formation, induced Wnt/β-catenin signaling, and polyp formation (He et al., 2007). This parallels the development of hamartomatous lesions found in Cowden disease patients, known to inherit PTEN mutations (Liaw et al., 1997). Interestingly, the involvement of the stroma in the development of tumors associated with deregulated Hedgehog, BMP, and PTEN signaling is underlined in multiple studies (Auclair, Benoit, Rivard, Mishina, & Perreault, 2007; Beppu et al., 2008; Büller et al., 2015; Gerling et al., 2016; He et al., 2007; Howe et al., 1998; Sneddon et al., 2006; van den Brink et al., 2004). These signaling pathways therefore simultaneously contribute to differentiation along the crypt-villus axis, determination of β-catenin active or inactive state despite a gradient of Wnt, and interactions with the crypt-adjacent mesenchyme.

3.3 Notch Signaling

The Notch pathway is also key to ISC maintenance and daughter cell-fate determination. Contrary to the previously described signaling pathways, Notch signaling works by lateral inhibition between two adjacent cells. Notch ligands Dll1, Dll4, or Jagged1 at the cell surface will bind the Notch1 or 2 receptors on neighboring cells (Sander & Powell, 2004; Schröder & Gossler, 2002). Upon ligand binding, the receptor is cleaved by ADAM10, shedding its extracellular portion, and by a γ-secretase complex, releasing its intracellular domain (NICD). The NICD translocates into the nucleus and, alongside cofactors like RBPJκ (recombination signal-binding protein for immunoglobulin κJ region), promotes the transcription of target genes involved in proliferation and differentiation, including Hes (hairy and enhancer of split)-family transcription factors. The best-characterized NICD target, *Hes1*, represses transcription of Notch ligands and of *Atoh1* (atonal

BHLH transcription factor 1, *Math1* in mice), both key to secretory lineage commitment, as well as that of genes encoding cyclin-dependent kinase inhibitors $p27^{Kip1}$ and $p57^{Kip2}$ (Milano et al., 2004; Obata et al., 2012; Stanger, Datar, Murtaugh, & Melton, 2005; Tsai et al., 2014; Yang, Bermingham, Finegold, & Zoghbi, 2001).

Signal inhibition all along the pathway by genetic deletion of *Dll1* and *Dll4* (Pellegrinet et al., 2011), *Notch1* and *Notch2* (Carulli, Keeley, Demitrack, Chung, & Samuelson, 2016; Riccio et al., 2008), *ADAM10* (Tsai et al., 2014), *RBPJκ* (van Es, Jay, et al., 2005; van Es, van Gijn, et al., 2005), or multiple *Hes* genes (Ueo et al., 2012) or by treatment with γ-secretase inhibitors (Milano et al., 2004; van Es, Jay, et al., 2005; van Es, van Gijn, et al., 2005; VanDussen et al., 2012) converts all proliferative cells in the crypt into postmitotic secretory cells. This includes $Lgr5^+$ ISC, which are consequently lost (Carulli et al., 2016; VanDussen et al., 2012), leading to lack of epithelial regeneration, nutrient malabsorption, weight loss, and death. Accordingly, lineage tracing experiments reveal active Notch signaling in ISC (Fre et al., 2011; Pellegrinet et al., 2011), and *Olfm4* was established as a direct target of the NICD and RBPJκ (VanDussen et al., 2012). On the other hand, expression of a constitutively active NCID results in increased proliferation and secretory cell depletion (Fre et al., 2005; Stanger et al., 2005). *Math1* deletion also results in depletion of all secretory lineages (Durand et al., 2012; Kim, Escudero, & Shivdasani, 2012; Yang et al., 2001). Notch signaling therefore maintains the proliferating ISC pool and suppresses secretory lineage engagement. Beyond this, however, Notch signaling does not appear necessary for absorptive lineage differentiation as simultaneous *Math1* deletion and Notch disruption does not prevent enterocyte differentiation (Kazanjian, Noah, Brown, Burkart, & Shroyer, 2010; Kim & Shivdasani, 2011; van Es, de Geest, van de Born, Clevers, & Hassan, 2010). Furthermore, while the effects of Notch activation on proliferation are only possible in Wnt-activated cells, its effects on differentiation are Wnt independent (Fre et al., 2009). The current model postulates that Wnt-induced stem cells are maintained in a Notch-high state by neighboring Notch ligand-expressing cells, and that upon exit from the ISC niche, the bimodal nature of the pathway will result in some daughters' expression of *Dll1* and *Math1* and repression of *Notch*, leading to their secretory lineage commitment. These cells will then repress such a commitment in their neighboring TA cells, ensuring an appropriate absorptive-to-secretory lineage ratio. Paneth cells express Notch ligands (Sato, van Es, et al., 2011) and thereby sustain active Notch signals in their neighboring

ISC, although other sources of Notch ligands also likely exist in the crypt as stem cells are maintained despite the absence of Paneth cells in the colon and in the small intestine of *Math1*-knockout mice (Durand et al., 2012; Kim et al., 2012).

Unsurprisingly, Notch signaling has been detected in both human and mouse intestinal adenomas, and aberrant Notch activation promotes adenoma formation in *Apc*-mutant mice (Fre et al., 2009; van Es, Jay, et al., 2005; van Es, van Gijn, et al., 2005). Conversely, γ-secretase inhibitor treatment partially converted adenoma cells of *Apc*-mutant mice into secretory cells, although this treatment is limited by its concurrent effects in the normal epithelium (van Es, Jay, et al., 2005; van Es, van Gijn, et al., 2005) and on other tissues. Furthermore, blocking Notch signaling does not prevent the initiation of Apc-mutant tumors, suggesting that β-catenin activation overrides the expected forced differentiation of proliferative cells (Peignon et al., 2011). Notably, Notch signaling is weaker in more advanced human carcinomas compared to adenomas, and its activation never leads to adenocarcinoma formation in *Apc*-mutant mice, indicating that Notch activation may be favorable (although not required) for tumor development but not for malignant progression (Fre et al., 2009).

4. ORGANOID CULTURE AND THE EXPERIMENTAL NICHE

Although in vitro culture of intestinal crypts has often been labeled as "niche independent" growth of ISC, it actually highlights the importance of a minimal molecular niche for proper ISC function. Among the key supplements to organoid culture medium are Wnt agonist and Lgr5 ligand R-spondin1, the BMP antagonist Noggin, and epidermal growth factor (EGF) (Sato et al., 2009). Little is currently known about the effects of EGF on ISC. In the adult intestine, EGF receptor (EGFR) is mainly expressed in the epithelium, especially in crypt CBC and TA cells (Yang et al., 2017), whereas Paneth cells highly express EGF ligands EFG and TGFα (transforming growth factor-α) (Sato, van Es, et al., 2011). EGFR-family receptor ErbB2 is also highly expressed on CBC (Sasaki et al., 2016), although its activity appears to be negatively regulated by Lrig1 (Wong et al., 2012). Production of another EGF-family ligand, Amphiregulin, from intestinal lymphocytes has recently been implicated in epithelial regeneration (Monticelli et al., 2015). Furthermore, EGFR signaling activates the RAS pathway, of which several components are constitutively activated in later-stage CRC.

When isolated ISC rather than whole crypts are cultured, exogenous Wnt ligands (typically Wnt3a) must be added to the culture medium. Although this allows for ISC maintenance and proliferation, the resulting organoids no longer contain all the differentiated cells of the homeostatic epithelium but rather seem to consist largely of undifferentiated progenitors that form a spheroid structure (Sato, van Es, et al., 2011). This is likely due to the ubiquitous delivery of Wnt ligand, as opposed to the more homeostatic gradient created by coculture with Paneth cells. The implication of different signaling pathways in intestinal homeostasis can be tackled in vitro as organoids recapitulate the in vivo epithelial response—a hyperplastic crypt might develop into an undifferentiated spheroid whereas stem cell failure will lead to organoid death. This system also has the added advantage of easier manipulation of treatments and the possibility to follow epithelial response over time while excluding mesenchymal response (Yin et al., 2014).

The supplemented growth factors described above apply for murine small intestinal crypts (without exogenous Wnt) or isolated stem cells (with added Wnt3a). Isolated colon crypts produce insufficient amounts of Wnt ligands to maintain colonic ISC in culture, therefore Wnt3a must be added for colonic organoid growth (Sato, Stange, et al., 2011). Like small intestinal organoids with exogenous Wnt, however, these do not properly differentiate. In the case of human small intestinal and colon crypts, addition of further factors is required for efficient long-term culture—gastrin, nicotinamide, an ALK receptor inhibitor, and an inhibitor of p38 MAPK (mitogen-activated protein kinase)—suggestive of additional important pathways in human ISC function. It is interesting to note that ISC presenting mutations in genes commonly mutated in CRC, or organoids derived from human colorectal tumors have decreasing needs for growth factors in the culture medium, suggesting a decreasing niche factor dependence in tumor stem cells (Drost et al., 2015; Fujii et al., 2016).

A final component that must not be neglected for ISC maintenance and organoid growth in vitro is the substrate upon which ISC are plated. In vivo, the epithelium is surrounded by an extracellular matrix (ECM) enriched in laminin at the crypt base. For organoid growth, ISC are embedded in laminin-rich matrigel. Isolated epithelial cells die by anoikis, a form of apoptosis following loss of matrix interactions, and the addition of ROCK signaling inhibitor Y-27632 to the culture medium also improves organoid formation, further highlighting the importance of matrix interactions (Sato et al., 2009). In vivo, the basal side of intestinal epithelial cells is in direct contact with a network of extracellular proteins that form the basement membrane (BM), including type IV collagen, laminin, and proteoglycans. BM components are produced both from epithelial and mesenchymal cells including subepithelial myofibroblasts. In the lamina propria, similar components including type I collagen and fibronectin form a three-dimensional ECM. Integrins recognize laminin, collagen, and fibronectin, inducing cytoplasmic signaling cascades that affect cytoskeleton remodeling, cellular growth, and cell cycle progression. Differential distributions of laminins and integrins have been observed along the crypt-villus axis, affecting not just adhesion, but also proliferation, migration, and sensitivity to anoikis of epithelial cells. BM and ECM components as well as integrin signaling have been implicated in CRC. Like EGF, integrins also induce the Ras signaling cascade, therefore the activating *KRAS* and *BRAF* mutations in later-stage CRC could promote anchorage-independent growth of tumor cells. Mmps are soluble or membrane-bound proteases produced both by the epithelium and mesenchyme that can remodel the BM and ECM, and several of these have also been implicated in CRC progression and inflammatory bowel disease. Lastly, the ECM can also sequester different growth factors, like transforming growth factor β (TGFβ), which become accessible by matrix remodeling or during epithelial damage (Hynes, 2009; Lussier, Basora, Bouatrouss, & Beaulieu, 2000; Murgia et al., 1998; Worthley, Giraud, & Wang, 2010).

5. THE EPITHELIAL NICHE: PANETH CELLS AND DEEP CRYPT SECRETORY CELLS

One epithelial cell type stands out with regard to its involvement in the ISC niche. Even as far back as the 1970s, Cheng and Leblond proposed a tight functional link between CBC and their position between Paneth cells. In link with their demarcating feature, large secretory granules filled with α-defensins (HD5 and HD6 in humans) and other AMP, Paneth cells were initially recognized for their role in innate immunity. Paneth cells normally release their granule contents into the crypt lumen in basal conditions and increasingly so in response to stimuli such as bacterial surface products and other TLR (Toll-like receptor) agonists (Ayabe et al., 2000; Vaishnava, Behrendt, Ismail, Eckmann, & Hooper, 2008) or proinflammatory cytokines like interferon gamma (Farin et al., 2014). Although α-defensins are constitutively expressed by Paneth cells under the control of the Wnt pathway and Tcf4 (Andreu et al., 2008; van Es, Jay, et al., 2005; van Es, van Gijn, et al., 2005), even in the absence of luminal microbes before birth or in germ-free conditions, other AMP are upregulated in the presence of bacteria (Ayabe et al., 2000; Harris, 1996; Pütsep et al., 2000). These AMP include lysozyme, secretory phospholipase A2, and Reg3 (regenerating islet-derived protein 3, Reg3A in humans, Reg3γ and Reg3β in mice). These secreted AMP not only protect the crypt epithelium from enteric pathogens (Vaishnava et al., 2008), but also directly affect the composition of the microbiota (Salzman et al., 2010). In this manner, Paneth cells regulate homeostatic interactions between the epithelium and the intestinal microbiota, which could have huge repercussions on the inflammatory, and therefore regenerative, state of the intestine.

Paneth cells also produce many of the essential trophic factors regulating the ISC niche, including Wnt3, Wnt6, Wnt9, Dll4, Dll1, EGF, and TGFα (Gregorieff et al., 2005; Sato, van Es, et al., 2011). This hints at a major role of these cells in the maintenance of ISC and regulation of their proliferation. On this basis, it has been proposed that the Paneth cell zone defines the ISC niche. As ISC divide and their numbers increase, they will neutrally compete to either remain between Paneth cells and therefore maintain an ISC identity or drift out of the niche and adopt a TA fate (Ritsma et al., 2014; Snippert et al., 2010). An opposing hierarchical model argues that ISC divide mostly asymmetrically, giving rise to both a new stem daughter and a TA daughter that exits the niche. Both models are compatible with early lineage tracing experiments and the observed drift toward monoclonality of ISC within a single crypt (Gordon et al., 1992). Consistent with Paneth cells delimiting the ISC niche, partial ablation of Paneth cells results in a decrease in the number of $Lgr5^+$ CBC, with the remaining CBC crowded around the remaining Paneth cells (Sato, van Es, et al., 2011). Surprisingly, the total elimination of Paneth cells does not result in the loss of ISC (Durand et al., 2012; Kim et al., 2012; Shroyer et al., 2007). The space normally occupied by Paneth cells is filled by normal amounts of CBC with active Wnt/β-catenin signaling. In fact, proliferation is increased in Paneth cell-deficient crypts (Durand et al., 2012; Kim et al., 2012; Shroyer et al., 2007) and the regenerative capacity of ISC following irradiation is not affected (Durand et al., 2012). Paneth cells are therefore not required for ISC survival and proliferation. They are also not essential for stem cell localization, as CBC are retained at the crypt bottom despite the lack of Paneth cells. Although this may be shocking in light of the importance attributed to Paneth cells, several lines of evidence comfort the dispensability of Paneth cells for ISC function. Not only is the colon devoid of Paneth cells, but the small intestines of certain mammalian species, namely dogs and pigs, also lack Paneth cells (Potten, Booth, & Pritchard, 1997).

In vitro, however, isolated Paneth cell-deficient crypts cannot form organoids (Durand et al., 2012). The same is true of Paneth cell-derived Wnt3, the expression of which is dispensable for stem cell maintenance in vivo but essential for the mesenchyme-independent growth of intestinal organoids in vitro (Farin et al., 2012). The absence of Paneth cells in vivo must therefore be compensated by other as-of-yet unknown niche cells likely providing compensatory Wnt in response to the lack of Paneth cell signals. Wnt3- or Paneth cell-deficient organoids can be maintained in culture by addition of exogenous Wnt3a or by coculture with intestinal mesenchymal cells (Durand et al., 2012; Farin et al., 2012; Kabiri et al., 2014; Sato, van Es, et al., 2011). Wnt2b is the only Wnt ligand normally expressed by the mesenchyme capable of rescuing Wnt3- or Paneth cell-deficient organoid growth; it has therefore been suggested as the potential compensatory Wnt signal in vivo (Farin et al., 2012). Of course, such a rescue has not been formally demonstrated in vivo, and the existence of another compensatory mechanism should not be excluded. Of note, exogenous Wnt6 or Wnt9b also rescue Wnt3-deficient organoid growth, suggesting that the amount produced by the crypt epithelial cells is not sufficient to compensate Wnt3 loss. Recently, Farin et al. further examined the establishment of the Wnt3 gradient from Paneth cells. Surprisingly, they found Wnt3 to act not as a diffusible gradient but as a Fzd-bound signal on the basolateral membrane of cells at the crypt bottom that is diluted up the crypt by cell division (Farin et al., 2016). They also report that Wnt3 does not localize in Paneth cell secretion granules, implying that it is secreted through another mechanism. This is in agreement with a previous report that put forward a Rab8a vesicle-dependent secretion of Wnt ligands (Das et al., 2015).

Paneth cells have been shown not just to provide a defined niche for ISC, but also to orchestrate their response to nutrient availability. It makes sense for the intestinal epithelium, whose principal function is nutrient absorption, to be highly responsive to diet and nutrient availability. In response to calorie restriction, villi shorten along with the TA compartment, while CBC numbers increase. This increase in ISC numbers and proliferation is accompanied by an increase in Paneth cells. Inhibition of mTORC1 (mammalian target of rapamycin complex 1) is a classically described cellular response to nutrient deprivation. In 2012, Yilmaz and coworkers showed that the augmented ISC renewal following calorie restriction is dependent on reduced mTORC1 activity in Paneth cells, resulting in increased extracellular cyclic ADP ribose production (Yilmaz et al., 2012). Igarashi and Guarente complemented this study by showing that this paracrine signal activates Sirtuin1 and mTORC1 signaling in ISC, promoting their self-renewal. Paneth cell signaling therefore overrides direct nutrient sensing in ISC to coordinate a proper response (Igarashi & Guarente, 2016). A new study by Rodriguez-Coman et al. showed that while Lgr5$^+$ CBC have high mitochondrial activity, Paneth cells show high glycolysis levels. Blocking the glycolytic activity of Paneth cells hindered their ability to support organoid formation in culture, while blocking oxidative phosphorylation in CBC also reduced organoid formation. On the other hand, providing lactate to Lgr5$^+$ CBC enhanced organoid formation. The authors therefore propose that Paneth cells, through their high glycolytic activity, provide lactate to CBC, enabling their high mitochondrial metabolism and proliferative capacity (Rodríguez-Colman et al., 2017). These studies put forward an entirely new niche function of Paneth cells: supporting ISC metabolism, and orchestrating their adaptation to their metabolic environment. Paradoxically, a high-fat diet also leads to an increase in ISC numbers along with decreased Paneth cell numbers (Beyaz et al., 2016). This is supposedly by a mechanism involving peroxisome proliferator-activated receptor δ (PPARδ) signaling in CBC conferring Paneth-independent growth in vitro.

Although Paneth cells are absent in the colon, Lgr5$^+$ CBC are still found intercalated between secretory cells in the colon mucosa. These cells, first coined "deep crypt secretory" (DCS) cells in 1983 by G.G. Altmann, are distinct from goblet cells but still contain mucous vacuoles in their cytoplasm. Morphologic analysis of DCS cells reveals many shared features with small intestine Paneth cells: they have a highly developed endoplasmic reticulum and Golgi complex for glycoprotein production, they mature as they migrate down from the midcrypt to the crypt base, and their numbers vary from the proximal to the distal colon (Altmann, 1983). With this background in mind, Rothenberg et al. identified a new goblet cell type that expresses several factors involved in ISC maintenance, such as Dll1, Dll4, and EGF. These cells can be recognized by marker cKit, also expressed by small intestinal Paneth cells, and intercalate between Lgr5$^+$ CBC in the colon (Rothenberg et al., 2012). These cells were also shown to express Reg4 (regenerating islet-derived 4) and correspond to Altmann's DCS cells. Just recently, Sasaki et al. characterized the gene expression profile of Reg4$^+$ DCS cells and found that they express several other Notch and EGF ligands along with goblet cell markers, although their signature matches that of

Paneth cells more closely than that of goblet cells. Specific ablation of DTR-expressing DCS cells upon DT treatment led to a loss of $Lgr5^+$ stem cells (Sasaki et al., 2016). Consistently, coculture of $Lgr5^+$ stem cells with $cKit^+$ and $Reg4^+$ DCS cells improves organoid growth while the disruption of DCS cells hinders it. Finally, DCS cell numbers are regulated by Notch signaling, as is the case for Paneth cells (Rothenberg et al., 2012; Sasaki et al., 2016). One key difference stands out between the two cell types: DCS cells, unlike Paneth cells, do not produce any Wnt ligands (Sasaki et al., 2016), although colon Lgr5+ cells do depend on them for growth in vitro (Sato, Stange, et al., 2011). As in the Paneth cell-deficient small intestine, another extraepithelial source of Wnt likely exists in the colon to maintain ISC homeostasis.

6. THE MESENCHYMAL NICHE

The lamina propria comprises various mesenchymal cell types, including fibroblasts and myofibroblasts, smooth muscle cells, vascular pericytes and endothelial cells, enteric neurons, and hematopoietic cells. Of course, decrypting the complex interactions between mesenchymal cells and the intestinal epithelium relies on ongoing characterization of the different subsets of stromal cells and their functions. Stromal cells have been shown to play a crucial role in homeostasis of the intestinal epithelium as well as in its response to different physiological stresses. Several key signaling pathways regulating epithelial proliferation and differentiation involve epithelial–mesenchymal communication, although the specific mesenchymal cell types involved are not yet fully established. These signals include the aforementioned Hedgehog and BMP signals, canonical and noncanonical Wnt ligands and Wnt antagonists, EGF ligands and R-spondins, as well as ECM components. Moreover, as small intestinal organoids lack the formation of extruding villi, the lamina propria is likely also necessary for complete crypt-villus axis patterning and morphology. Importantly, stromal cells have additionally been shown to contribute to intestinal tumor development and progression (Powell, Pinchuk, Saada, Chen, & Mifflin, 2011).

As previously noted, stromal cells provide Wnt signals both in the colon, where epithelial niche cells do not produce Wnt ligands, and in the small intestine, in complement to Paneth cells. Similar to the epithelial deletion of Wnt3 by Farin et al. or Paneth cell ablation by Durand et al., Kabiri and coworkers used epithelium-specific genetic ablation of *Porcn*, encoding the ER O-acetlytransferase Porcupine required for Wnt secretion and activity, to demonstrate that epithelial Wnts were dispensable in vivo but necessary for organoid growth in vitro (Kabiri et al., 2014). On the other hand, pharmacologic inhibition of Porcupine (affecting both the epithelium and surrounding mesenchyme) or whole-body deletion of Wntless (another protein required for Wnt secretion) results in $Lgr5^+$ ISC loss (Kabiri et al., 2014; Valenta et al., 2016). Kabiri et al. further demonstrate that Porcupine-deficient organoid growth can be rescued by coculture with intestinal stromal cells (Kabiri et al., 2014), proving the compensatory ability of the intestinal stroma in this context. Several stromal cell types located adjacent to the epithelium, including subepithelial myofibroblasts, secrete Wnt2b and Wnt5a (Valenta et al., 2016). The fact that simultaneous deletion of Porcupine in both epithelial cells and myofibroblasts still does not disrupt crypt homeostasis suggests that other Wnt-producing cells of the lamina propria balance out the lack of Wnt ligand from these cells (San Roman, Jayewickreme, Murtaugh, & Shivdasani, 2014). A new report focused on $CD34^+$ $gp38^+$ mesenchymal cells, which are negative for endothelial (CD31), hematopoietic (CD45), and myofibroblast (α-smooth muscle actin) markers and surround crypts in both the small intestine and colon. Coculture of organoids with these cells results in the formation of highly proliferative spheroids, similar to addition of exogenous Wnt. Moreover, this overrules the need for R-spondin supplementation. Indeed, $CD34^+$ $gp38^+$ crypt stromal cells express high levels of *Wnt2b*, *Gremlin1*, and *Rspo1* (encoding R-spondin1) as compared to other lamina propria mesenchymal cell types (Stzepourginski et al., 2016). These cells could therefore be an essential mesenchymal component of the ISC niche. In line with this, depletion of mesenchymal cells expressing Foxl1 (forkhead box L1, previously known as Fkh6), a transcription factor notably expressed in CD34+ gp38+ cells, results in disrupted epithelial morphology including loss of stem and proliferative cells (Aoki et al., 2016).

Immune cells also reside in the lamina propria and can interact with the epithelium and its stem cells. Defined subsets of hematopoietic cells such as γδ T lymphocytes (Komano et al., 1995) or macrophages (Bansal, Trinath, Chakravortty, Patil, & Balaji, 2011) have been reported to affect epithelial homeostasis or recovery following infection or tissue damage, although the signals involved in these interactions remain unclear. Coculture of intestinal organoids with isolated lamina propria lymphocytes stimulates $Lgr5^+$ CBC expansion and growth of the organoids in a way that is dependent on interleukin (IL)-22 production. The same effect is recapitulated both with isolated group 3 innate lymphoid cells (ILCs) and with recombinant IL-22 in a dose-dependent manner. This process is independent of Paneth cells or altered Wnt, Notch, or EGF activity, instead relying on Stat3 signaling in ISC (Lindemans et al., 2015). Upon injury in vivo, crypt-adjacent ILCs, which lack antigen receptors but respond to IL-23, produce IL-22 while epithelial cells upregulate their expression of the IL-22 receptor, overall promoting epithelial regeneration (Hanash et al., 2012; Lindemans et al., 2015). In counterpart, exacerbated IL-22 signals can promote tumorigenesis in the intestine (Huber et al., 2012; Kirchberger et al., 2013), emphasizing its effects on ISC.

7. THE MICROBIAL NICHE

Opposite to the lamina propria, the intestinal epithelium is exposed to the contents of the intestinal leumen, incluing passing digested foods but also the resident microbiota. In germ-free mice and rats, epithelial kinetics are slowed (Abrams, Bauer, & Sprinz, 1963; Alam, Midtvedt, & Uribe, 1994; Reikvam et al., 2011), suggesting a direct or indirect effect of the microbiota on ISC dynamics. Aside from an indirect effect on immune cell-epithelium cross-talks, Jones et al. identified one mechanism through which commensal bacteria could affect ISC. Intestinal epithelial cells express NADPH oxidases like Nox1 that produce reactive oxygen species (ROS) as an antimicrobial defense mechanism. Well-regulated levels of ROS can stimulate proliferation of stem and progenitor cells (Jones et al., 2013; Myant et al., 2013). Specific bacterial species (including *Lactobacillus rhamnosis* but not *Escherichia coli*) induce epithelial ROS and subsequent proliferation in a Nox1-dependent manner (Jones et al., 2013). Later, Jones et al. further showed a cytoprotective effect of *Lactobacillus* on crypt cells in the context of irradiation via the activation of an antioxidant response mediated by Nrf2 (nuclear factor erythroid 2) (Jones et al., 2015). Another mechanism by which the microbiota affects ISC is through microbial metabolites, several of which were discovered to suppress colon stem and progenitor cell proliferation in a recent screen by Stappenbeck's group (Kaiko et al., 2016). Of these, butyrate, a product of bacterial dietary fiber fermentation, had the most drastic effects, suppressing proliferation at low concentrations and inducing apoptosis at concentrations closer to those found in the colon lumen. The authors postulate that the crypt architecture and colonocytes' ability to oxidize and metabolize butyrate create a luminal gradient of butyrate down the crypt regulating stem and progenitor cell proliferation. Interestingly, butyrate-mediated inhibition of proliferation depends on epigenetic changes involving histone deacetylase 3 (HDAC3), previously reported to link bacterial signals to epithelial response in the intestine (Alenghat et al., 2013).

Because of their localization furthest from luminal contents in the mucus- and AMP-protected crypt bottom, it was long thought that homeostatic ISC benefited from a sterile environment. Challenging this notion, work by Sansonetti's group revealed the existence of a restricted set of bacteria in colonic crypts of healthy mice, which they termed the "crypt-specific core microbiota" (Pédron et al., 2012). Interestingly, this predominantly aerobic population resembles the microbiota found in the midgut of invertebrates, hinting at a potential coevolutionary selection of commensals favoring proper gut dynamics. Host cells can recognize microbial signals via pattern recognition receptors (PRR), including transmembrane TLR and cytoplasmic Nod-like receptors (NLR). The stimulation of PRR by their corresponding antigens induce a signaling cascade via their common downstream adaptor molecule MyD88 (myeloid differentiation primary response gene 88), often resulting in induction of MAPK and NF-κB (nuclear factor κ-light-chain-enhancer of activated B cells) signaling and production of proinflammatory cytokines. Both the commensal microbiota and MyD88 activity are required for epithelial recovery from DSS-induced colitis, revealing a surprising regenerative effect of microbiota-induced PRR signaling in the gut (Rakoff-Nahoum, Pglino, Eslami-Varzaneh, Edberg, & Medzhitov, 2004). MyD88 signaling also contributes to intestinal tumorigenesis (Rakoff-Nahoum & Medzhitov, 2007). These observations were made all the more relevant by reports of direct expression of PRR TLR4 and Nod2 (nucleotide-binding oligomerization domain-containing protein 2) by $Lgr5^+$ ISC directly. The stimulation of these PRR in vivo or in organoid cultures has drastically different effects on ISC; whereas treatment with TLR4 antigen lipopolysaccharide induces apoptosis of $Lgr5^+$ ISC (Neal et al., 2012), treatment with the Nod2 agonist muramyl dipeptide protects them from regenerative or ROS-induced stress (Nigro, Rossi, Commere, Jay, & Sansonetti, 2014). Stem cells are therefore in direct communication with luminal microbes and can thereby adjust their response to the luminal microenvironment. These studies also bring a new perspective on the reported link between the presence of particular commensals and pathogens and the development of CRC (Castellarin et al., 2011; Dejea et al., 2014; Kostic et al., 2012, 2013; Machida-Montani et al., 2007).

Through its unique organization and multifaceted microenvironment, the intestinal system truly expands the boundaries of what is classically called the stem cell niche. Taking into account not just epithelial signals but also those from the mesenchyme and lumen, ISC are able to continuously regenerate an epithelium adapted to the microenvironment they are faced with. Despite the continuous and rapid regeneration of this epithelium, cross-talk and redundancy between ISC niche components allow for a tightly controlled, robust system, and prevent unrestrained proliferation and tumorigenesis. Understanding the intricate regulatory networks behind this is made considerably easier with the development of modern techniques of genetic manipulation and ex vivo culture that will no doubt lead to great advances in the field in the years to come.

REFERENCES

Abrams, G., Bauer, H., & Sprinz, H. (1963). Influence of the normal flora on mucosal morphology and cellular renewal in the ileum. A comparison of germ-free and conventional mice. *Laboratory Investigation*, *12*, 355–364.

Alam, M., Midtvedt, T., & Uribe, A. (1994). Differential cell kinetics in the ileum and colon of germfree rats. *Scandinavian Journal of Gastroenterology*, *29*(5), 445–451.

Alenghat, T., Osborne, L. C., Saenz, S. A., Kobuley, D., Ziegler, C. G., Mullican, S. E., et al. (2013). Histone deacetylase 3 coordinates commensal-bacteria-dependent intestinal homeostasis. *Nature*, *504*(7478), 153–157. https://doi.org/10.1038/nature12687.

Altmann, G. G. (1983). Morphological observations on mucus-secreting nongoblet cells in the deep crypts of the rat ascending colon. *The American Journal of Anatomy*, *167*(1), 95–117. https://doi.org/10.1002/aja.1001670109.

Andreu, P., Colnot, S., Godard, C., Gad, S., Chafey, P., Niwa-Kawakita, M., et al. (2005). Crypt-restricted proliferation and commitment to the Paneth cell lineage following Apc loss in the mouse intestine. *Development*, *132*, 1443–1451. https://doi.org/10.1242/dev.01700.

Andreu, P., Peignon, G., Slomianny, C., Taketo, M. M., Colnot, S., Robine, S., et al. (2008). A genetic study of the role of the Wnt/β-catenin signalling in Paneth cell differentiation. *Developmental Biology*, *324*, 288–296. https://doi.org/10.1016/j.ydbio.2008.09.027.

Aoki, R., Shoshkes-Carmel, M., Gao, N., Shin, S., May, C. L., Golson, M. L., et al. (2016). Foxl1-expressing mesenchymal cells constitute the intestinal stem cell niche. *Cellular and Molecular Gastroenterology and Hepatology*, *2*(2), 175–188. https://doi.org/10.1016/j.jcmgh.2015.12.004.

Auclair, B. A., Benoit, Y. D., Rivard, N., Mishina, Y., & Perreault, N. (2007). Bone morphogenetic protein signaling is essential for terminal differentiation of the intestinal secretory cell lineage. *Gastroenterology*, *133*(3), 887–896. https://doi.org/10.1053/j.gastro.2007.06.066.

Ayabe, T., Satchell, D. P., Wilson, C. L., Parks, W. C., Selsted, M. E., & Ouellette, A. J. (2000). Secretion of microbicidal alpha-defensins by intestinal Paneth cells in response to bacteria. *Nature Immunology*, *1*(2), 113–118. https://doi.org/10.1038/77783.

Bansal, K., Trinath, J., Chakravortty, D., Patil, S. A., & Balaji, K. N. (2011). Pathogen-specific TLR2 protein activation programs macrophages to induce Wnt-beta-catenin signaling. *Journal of Biological Chemistry*, *286*(42), 37032–37044. https://doi.org/10.1074/jbc.M111.260414.

Barker, N., Ridgway, R. A., van Es, J. H., van de Wetering, M., Begthel, H., van den Born, M., et al. (2009). Crypt stem cells as the cells-of-origin of intestinal cancer. *Nature*, *457*(7229), 608–611. https://doi.org/10.1038/nature07602.

Barker, N., van Es, J. H., Kuipers, J., Kujala, P., van den Born, M., Cozijnsen, M., et al. (2007). Identification of stem cells in small intestine and colon by marker gene Lgr5. *Nature*, *449*(7165), 1003–1007. https://doi.org/10.1038/nature06196.

Batlle, E., Bacani, J., Begthel, H., Jonkheer, S., Gregorieff, A., van de Born, M., et al. (2005). Corrigendum: EphB receptor activity suppresses colorectal cancer progression. *Nature*, *436*(7052), 881. https://doi.org/10.1038/nature03941.

Batlle, E., Henderson, J. T., Beghtel, H., Van den Born, M. M. W., Sancho, E., Huls, G., et al. (2002). β-Catenin and TCF mediate cell positioning in the intestinal epithelium by controlling the expression of EphB/EphrinB. *Cell*, *111*(2), 251–263. https://doi.org/10.1016/S0092-8674(02)01015-2.

Beppu, H., Mwizerwa, O. N., Beppu, Y., Dattwyler, M. P., Lauwers, G. Y., Bloch, K. D., et al. (2008). Stromal inactivation of BMPRII leads to colorectal epithelial overgrowth and polyp formation. *Oncogene*, *27*(8), 1063–1070. https://doi.org/10.1038/sj.onc.1210720.

Beyaz, S., Mana, M. D., Roper, J., Kedrin, D., Saadatpour, A., Hong, S., et al. (2016). High-fat diet enhances stemness and tumorigenicity of intestinal progenitors. *Nature*, *531*(7592), 53–58. https://doi.org/10.1038/nature17173.

Bjerknes, M., & Cheng, H. (1981a). The stem-cell zone of the small intestinal epithelium. I. Evidence from Paneth cells in the adult mouse. *The American Journal of Anatomy*, *160*(1), 51–63.

Bjerknes, M., & Cheng, H. (1981b). The stem-cell zone of the small intestinal epithelium. III. Evidence from columnar, enteroendocrine, and mucous cells in the adult mouse. *The American Journal of Anatomy*, *160*(1), 77–91.

Bjerknes, M., & Cheng, H. (1999). Clonal analysis of mouse intestinal epithelial progenitors. *Gastroenterology*, *116*(1), 7–14.

Bjerknes, M., & Cheng, H. (2002). Multipotential stem cells in adult mouse gastric epithelium. *American Journal of Physiology Gastrointestinal and Liver Physiology*, *283*(3), G767–G777. https://doi.org/10.1152/ajpgi.00415.2001.

Bondow, B. J., Faber, M. L., Wojta, K. J., Walker, E. M., & Battle, M. A. (2012). E-cadherin is required for intestinal morphogenesis in the mouse. *Developmental Biology*, *371*(1), 1–12. https://doi.org/10.1016/j.ydbio.2012.06.005.

Breault, D. T., Min, I. M., Carlone, D. L., Farilla, L. G., Ambruzs, D. M., Henderson, D. E., et al. (2008). Generation of mTert-GFP mice as a model to identify and study tissue progenitor cells. *Proceedings of the National Academy of Sciences of the United States of America*, *105*(30), 10420–10425. https://doi.org/10.1073/pnas.0804800105.

Buczacki, S. J. a., Zecchini, H. I., Nicholson, A. M., Russell, R., Vermeulen, L., Kemp, R., et al. (2013). Intestinal label-retaining cells are secretory precursors expressing Lgr5. *Nature*, *495*(7439), 65–69. https://doi.org/10.1038/nature11965.

Büller, N. V. J. A., Rosekrans, S. L., Metcalfe, C., Heijmans, J., Van Dop, W. A., Fessler, E., et al. (2015). Stromal Indian hedgehog signaling is required for intestinal adenoma formation in mice. *Gastroenterology*, *148*(1), 170–180.e6. https://doi.org/10.1053/j.gastro.2014.10.006.

Carulli, A. J., Keeley, T. M., Demitrack, E. S., Chung, J., & Samuelson, L. C. (2016). Notch receptor regulation of intestinal stem cell homeostasis and crypt regeneration. *Developmental Biology*, *402*(1), 98–108. https://doi.org/10.1016/j.ydbio.2015.03.012. Notch.

Castellarin, M., Warren, R. L., Freeman, J. D., Warren, L., Freeman, J. D., Dreolini, L., et al. (2011). *Fusobacterium nucleatum* infection is prevalent in human colorectal carcinoma. *Genome Research*, *22*, 299–306. https://doi.org/10.1101/gr.126516.111.

Cheng, H. (1974). Origin, differentiation and renewal of the four main epithelial cell types in the mouse small intestine. II. Mucous cells. *The American Journal of Anatomy*, *141*(4), 481–501.

Cheng, H., & Leblond, C. P. (1974a). Origin, differentiation and renewal of the four main epithelial cell types in the mouse small intestine. V. Unitarian theory of the origin of the four epithelial cell types. *The American Journal of Anatomy*, *141*(4), 537–561.
Cheng, H., & Leblond, C. P. (1974b). Origin, differentiation and renewal of the four main epithelial cell types in the mouse small intestine. I. Columnar cell. *The American Journal of Anatomy*, *141*(4), 461–479.
Choi, H. J., Huber, A. H., & Weis, W. I. (2006). Thermodynamics of β-catenin–ligand interactions: The roles of the N- and C-terminal tails in modulating binding affinity. *Journal of Biological Chemistry*, *281*(2), 1027–1038. https://doi.org/10.1074/jbc.M511338200.
Colnot, S., Niwa-Kawakita, M., Hamard, G., Godard, C., Le Plenier, S., Houbron, C., et al. (2004). Colorectal cancers in a new mouse model of familial adenomatous polyposis: Influence of genetic and environmental modifiers. *Laboratory Investigation*, *84*, 1619–1630. https://doi.org/10.1038/labinvest.3700180.
Cortina, C., Palomo-Ponce, S., Iglesias, M., Fernández-Masip, J. L., Vivancos, A., Whissell, G., et al. (2007). EphB–ephrin-B interactions suppress colorectal cancer progression by compartmentalizing tumor cells. *Nature Genetics*, *39*(11), 1376–1383. https://doi.org/10.1038/ng.2007.11.
Crawford, H. C., Fingleton, B. M., Rudolph-Owen, L. A., Goss, K. J. H., Rubinfeld, B., Polakis, P., et al. (1999). The metalloproteinase matrilysin is a target of β-catenin transactivation in intestinal tumors. *Oncogene*, *18*(18), 2883–2891. https://doi.org/10.1038/sj.onc.1202627.
Darmoul, D., & Ouellette, A. J. (1996). Positional specificity of defensin gene expression reveals Paneth cell heterogeneity in mouse small intestine. *American Journal of Physiology*, *271*(1), G68–74.
Das, S., Yu, S., Sakamori, R., Vedula, P., Feng, Q., Flores, J., et al. (2015). Rab8a vesicles regulate Wnt ligand delivery and Paneth cell maturation at the intestinal stem cell niche. *Development*, *142*(12), 2147–2162. https://doi.org/10.1242/dev.121046.
Davis, H., Irshad, S., Bansal, M., Rafferty, H., Boitsova, T., Bardella, C., et al. (2015). Aberrant epithelial GREM1 expression initiates colonic tumorigenesis from cells outside the stem cell niche. *Nature Medicine*, *21*(1), 62–70. https://doi.org/10.1038/nm.3750.
de Lau, W., Barker, N., Low, T. Y., Koo, B.-K., Li, V. S. W., Teunissen, H., et al. (2011). Lgr5 homologues associate with Wnt receptors and mediate R-spondin signalling. *Nature*, *476*(7360), 293–297. https://doi.org/10.1038/nature10337.
Dejea, C. M., Wick, E. C., Hechenbleikner, E. M., White, J. R., Mark Welch, J. L., Rossetti, B. J., et al. (2014). Microbiota organization is a distinct feature of proximal colorectal cancers. *Proceedings of the National Academy of Sciences of the United States of America*, *111*(51), 18321–18326. https://doi.org/10.1073/pnas.1406199111.
Dow, L. E., O'Rourke, K. P., Simon, J., Tschaharganeh, D. F., Van Es, J. H., Clevers, H., et al. (2015). Apc restoration promotes cellular differentiation and reestablishes crypt homeostasis in colorectal cancer. *Cell*, *161*, 1539–1552. https://doi.org/10.1016/j.cell.2015.05.033.
Drost, J., van Jaarsveld, R. H., Ponsioen, B., Zimberlin, C., van Boxtel, R., Buijs, A., et al. (2015). Sequential cancer mutations in cultured human intestinal stem cells. *Nature*, *521*(7550), 43–47. https://doi.org/10.1038/nature14415.
Durand, A., Donahue, B., Peignon, G., Letourneur, F., Cagnard, N., Slomianny, C., et al. (2012). Functional intestinal stem cells after Paneth cell ablation induced by the loss of transcription factor Math1 (Atoh1). *Proceedings of the National Academy of Sciences of the United States of America*, *109*(23), 8965–8970. https://doi.org/10.1073/pnas.1201652109.
Fan, L. F., Dong, W. G., Jiang, C. Q., Xia, D., Liao, F., & Yu, Q. F. (2010). Expression of putative stem cell genes Musashi-1 and beta1-integrin in human colorectal adenomas and adenocarcinomas. *International Journal of Colorectal Disease*, *25*(1), 17–23. https://doi.org/10.1007/s00384-009-0791-2.

Farin, H. F., Jordens, I., Mosa, M. H., Basak, O., Korving, J., Tauriello, D. V. F., et al. (2016). Visualization of a short-range Wnt gradient in the intestinal stem-cell niche. *Nature, 530*(7590), 340–343. https://doi.org/10.1038/nature16937.

Farin, H. F., Karthaus, W. R., Kujala, P., Rakhshandehroo, M., Schwank, G., Vries, R. G. J., et al. (2014). Paneth cell extrusion and release of antimicrobial products is directly controlled by immune cell-derived IFN-γ. *The Journal of Experimental Medicine, 211*(7), 1393–1405. https://doi.org/10.1084/jem.20130753.

Farin, H. F., Van Es, J. H., & Clevers, H. (2012). Redundant sources of Wnt regulate intestinal stem cells and promote formation of Paneth cells. *Gastroenterology, 143*(6), 1518–1529.e7. https://doi.org/10.1053/j.gastro.2012.08.031.

Fevr, T., Robine, S., Louvard, D., & Huelsken, J. (2007). Wnt/beta-catenin is essential for intestinal homeostasis and maintenance of intestinal stem cells. *Molecular and Cellular Biology, 27*(21), 7551–7559. https://doi.org/10.1128/MCB.01034-07.

Fodde, R., Fodde, R., Smits, R., Smits, R., Clevers, H., & Clevers, H. (2001). APC, signal transduction and genetic instability in colorectal cancer. *Nature Reviews Cancer, 1*(1), 55–67. https://doi.org/10.1038/35094067.

Fre, S., Hannezo, E., Sale, S., Huyghe, M., Lafkas, D., Kissel, H., et al. (2011). Notch lineages and activity in intestinal stem cells determined by a new set of knock-in mice. *PLoS One, 6*(10), e25785. https://doi.org/10.1371/journal.pone.0025785.

Fre, S., Huyghe, M., Mourikis, P., Robine, S., Louvard, D., & Artavanis-Tsakonas, S. (2005). Notch signals control the fate of immature progenitor cells in the intestine. *Nature, 435*(7044), 964–968. https://doi.org/10.1038/nature03589.

Fre, S., Pallavi, S. K., Huyghe, M., Laé, M., Janssen, K.-P., Robine, S., et al. (2009). Notch and Wnt signals cooperatively control cell proliferation and tumorigenesis in the intestine. *Proceedings of the National Academy of Sciences of the United States of America, 106*(15), 6309–6314. https://doi.org/10.1073/pnas.0900427106.

Fujii, M., Shimokawa, M., Date, S., Takano, A., Matano, M., Nanki, K., et al. (2016). A colorectal tumor organoid library demonstrates progressive loss of niche factor requirements during tumorigenesis. *Cell Stem Cell, 18*(6), 827–838. https://doi.org/10.1016/j.stem.2016.04.003.

Fukuda, M., Mizutani, T., & Mochizuki, W. (2014). Small intestinal stem cell identity is maintained with functional Paneth cells in heterotopically grafted epithelium onto the colon. *Genes & Development, 28*, 1752–1757. https://doi.org/10.1101/gad.245233.114.

Gerbe, F., Van Es, J. H., Makrini, L., Brulin, B., Mellitzer, G., Robine, S., et al. (2011). Distinct ATOH1 and Neurog3 requirements define tuft cells as a new secretory cell type in the intestinal epithelium. *Journal of Cell Biology, 192*(5), 767–780. https://doi.org/10.1083/jcb.201010127.

Gerling, M., Büller, N. V. J. A., Kirn, L. M., Joost, S., Frings, O., Englert, B., et al. (2016). Stromal Hedgehog signalling is downregulated in colon cancer and its restoration restrains tumour growth. *Nature Communications, 7*, 12321. https://doi.org/10.1038/ncomms12321.

Ginjala, V., Nacerddine, K., Kulkarni, A., Oza, J., Hill, S. J., Yao, M., et al. (2011). BMI1 is recruited to DNA breaks and contributes to DNA damage-induced H2A ubiquitination and repair. *Molecular and Cellular Biology, 31*(10), 1972–1982. https://doi.org/10.1128/MCB.00981-10.

Gordon, J. I. (1989). Intestinal epithelial differentiation: New insights from chimeric and transgenic mice. *The Journal of Cell Biology, 108*(4), 1187–1194. Retrieved from http://www.pubmedcentral.nih.gov/articlerender.fcgi?artid=2115507&tool=pmcentrez&rendertype=abstract.

Gordon, J. I., Schmidt, G. H., & Roth, K. A. (1992). Studies of intestinal stem cells using normal, chimeric, and transgenic mice. *The FASEB Journal, 6*(12), 3039–3050. Retrieved from http://www.nature.com/doifinder/10.1038/cr.2017.59.

Graham, T. A., Weaver, C., Mao, F., Kimelman, D., & Xu, W. (2000). Crystal structure of a β-catenin/Tcf complex. *Cell, 103*(6), 885–896. https://doi.org/10.1016/S0092-8674(00)00192-6.

Gregorieff, A., Pinto, D., Begthel, H., Destrée, O., Kielman, M., & Clevers, H. (2005). Expression pattern of Wnt signaling components in the adult intestine. *Gastroenterology, 129*(2), 626–638. https://doi.org/10.1016/j.gastro.2005.06.007.

Grün, D., Lyubimova, A., Kester, L., Wiebrands, K., Basak, O., Sasaki, N., et al. (2015). Single-cell messenger RNA sequencing reveals rare intestinal cell types. *Nature, 525*(7568), 251–255. https://doi.org/10.1038/nature14966.

Hanash, A. M., Dudakov, J. A., Hua, G., O'Connor, M. H., Young, L. F., Singer, N. V., et al. (2012). Interleukin-22 protects intestinal stem cells from immune-mediated tissue damage and regulates sensitivity to graft versus host disease. *Immunity, 37*(2), 339–350. https://doi.org/10.1016/j.immuni.2012.05.028.

Hao, H.-X., Xie, Y., Zhang, Y., Charlat, O., Oster, E., Avello, M., et al. (2012). ZNRF3 promotes Wnt receptor turnover in an R-spondin-sensitive manner. *Nature, 485*(7397), 195–200. https://doi.org/10.1038/nature11019.

Harada, N., Tamai, Y., Ishikawa, T. O., Sauer, B., Takaku, K., Oshima, M., et al. (1999). Intestinal polyposis in mice with a dominant stable mutation of the β-catenin gene. *The EMBO Journal, 18*(21), 5931–5942. https://doi.org/10.1093/emboj/18.21.5931.

Haramis, A.-P. G., Begthel, H., van den Born, M., van Es, J., Jonkheer, S., Offerhaus, G. J. A., et al. (2004). De novo crypt formation and juvenile polyposis on BMP inhibition in mouse intestine. *Science (New York, NY), 303*(5664), 1684–1686. https://doi.org/10.1126/science.1093587.

Harris, A. (1996). Human enteric defensins. *Journal of Biological Chemistry, 271*(8), 4038–4045. https://doi.org/10.1074/jbc.271.8.4038.

He, X. C., Yin, T., Grindley, J. C., Tian, Q., Sato, T., Tao, W. A., et al. (2007). PTEN-deficient intestinal stem cells initiate intestinal polyposis. *Nature Genetics, 39*(2), 189–198. https://doi.org/10.1038/ng1928.

He, X. C., Zhang, J., Tong, W.-G., Tawfik, O., Ross, J., Scoville, D. H., et al. (2004). BMP signaling inhibits intestinal stem cell self-renewal through suppression of Wnt–β-catenin signaling. *Nature Genetics, 36*(10), 1117–1121. https://doi.org/10.1038/ng1430.

Hermiston, M. L., Wong, M. H., & Gordon, J. I. (1996). Forced expression of E-cadherin in the mouse intestinal epithelium slows cell migration and provides evidence for non-autonomous regulation of cell fate in a self-renewing system. *Genes and Development, 10*, 985–996. https://doi.org/10.1101/gad.10.8.985.

Howe, J. R., Ringold, J. C., Summers, R. W., Mitros, F. A., Nishimura, D. Y., & Stone, E. M. (1998). A gene for familial juvenile polyposis maps to chromosome 18q21.1. *The American Journal of Human Genetics, 62*(5), 1129–1136. https://doi.org/10.1086/301840.

Huber, S., Gagliani1, N., Zenewicz, L. A., Huber, F. J., Bosurgi, L., Hu, B., et al. (2012). IL-22BP is regulated by the inflammasome and modulates tumorigenesis in the intestine. *Inflammatory Bowel Disease Monitor, 13*(3), 108–109. https://doi.org/10.1038/nature11535.

Huber, A. H., & Weis, W. I. (2001). The structure of the B-catenin/E-cadherin complex and the molecular basis of diverse ligand recognition by B-catenin. *Cell, 105*, 391–402. https://doi.org/10.1016/S0092-8674(01)00330-0.

Huels, D. J., Ridgway, R. A., Radulescu, S., Leushacke, M., Campbell, A. D., Biswas, S., et al. (2015). E-cadherin can limit the transforming properties of activating β-catenin mutations. *The EMBO Journal, 34*(18), 2321–2333. https://doi.org/10.15252/embj.201591739.

Hynes, R. O. (2009). The extracellular matrix: Not just pretty fibrils. *Science, 326*(5957), 1216–1219. https://doi.org/10.1126/science.1176009.

Igarashi, M., & Guarente, L. (2016). mTORC1 and SIRT1 cooperate to foster expansion of gut adult stem cells during calorie restriction. *Cell*, *166*(2), 436–450. https://doi.org/10.1016/j.cell.2016.05.044.

Ireland, H., Houghton, C., Howard, L., & Winton, D. J. (2005). Cellular inheritance of a Cre-activated reporter gene to determine Paneth cell longevity in the murine small intestine. *Developmental Dynamics*, *233*(4), 1332–1336. https://doi.org/10.1002/dvdy.20446.

Ireland, H., Kemp, R., Houghton, C., Howard, L., Clarke, A. R., Sansom, O. J., et al. (2004). Inducible Cre-mediated control of gene expression in the murine gastrointestinal tract: Effect of loss of β-catenin. *Gastroenterology*, *126*(5), 1236–1246. https://doi.org/10.1053/j.gastro.2004.03.020.

Ismail, H., Andrin, C., McDonald, D., & Hendzel, M. J. (2010). BMI1-mediated histone ubiquitylation promotes DNA double-strand break repair. *Journal of Cell Biology*, *191*(1), 45–60. https://doi.org/10.1083/jcb.201003034.

Itzkovitz, S., Lyubimova, A., Blat, I., Maynard, M., Van Es, J., Lees, J., et al. (2012). Single molecule transcript counting of stem cell markers in the mouse intestine. *Nature Cell Biology*, *14*(1), 106–114. https://doi.org/10.1038/ncb2384.Single.

Jadhav, U., Saxena, M., O'Neill, N. K., Saadatpour, A., Yuan, G. C., et al. (2016). Dynamic reorganization of chromatin accessibility signatures during dedifferentiation of secretory precursors into Lgr5+ intestinal stem cells. *Cell Stem Cell*, *21*(1), 65–77.

Jang, B. G., Kim, H. S., Kim, K. J., Rhee, Y.-Y., Kim, W. H., & Kang, G. H. (2016). Distribution of intestinal stem cell markers in colorectal precancerous lesions. *Histopathology*, *68*(4), 567–577. https://doi.org/10.1111/his.12787.

Jones, R. M., Desai, C., Darby, T. M., Luo, L., Wolfarth, A. A., Scharer, C. D., et al. (2015). Lactobacilli modulate epithelial cytoprotection through the Nrf2 pathway. *Cell Reports*, *12*, 1217–1225. https://doi.org/10.1016/j.celrep.2015.07.042.

Jones, R. M., Luo, L., Ardita, C. S., Richardson, A. N., Kwon, Y. M., Mercante, J. W., et al. (2013). Symbiotic lactobacilli stimulate gut epithelial proliferation via Nox-mediated generation of reactive oxygen species. *The EMBO Journal*, *32*(23), 3017–3028. https://doi.org/10.1038/emboj.2013.224.

Jubb, A. M., Chalasani, S., Frantz, G. D., Smits, R., Grabsch, H. I., Kavi, V., et al. (2006). Achaete-scute like 2 (ascl2) is a target of Wnt signalling and is upregulated in intestinal neoplasia. *Oncogene*, *25*(24), 3445–3457. https://doi.org/10.1038/sj.onc.1209382.

Kabiri, Z., Greicius, G., Madan, B., Biechele, S., Zhong, Z., Zaribafzadeh, H., et al. (2014). Stroma provides an intestinal stem cell niche in the absence of epithelial Wnts. *Development*, *141*, 2206–2215. https://doi.org/10.1242/dev.104976.

Kaiko, G. E., Ryu, S. H., Koues, O. I., Collins, P. L., Solnica-Krezel, L., Pearce, E. J., et al. (2016). The colonic crypt protects stem cells from microbiota-derived metabolites. *Cell*, *167*(4), 1137. https://doi.org/10.1016/j.cell.2016.10.034.

Kazanjian, A., Noah, T., Brown, D., Burkart, J., & Shroyer, N. F. (2010). Atonal homolog 1 is required for growth and differentiation effects of Notch/γ-secretase inhibitors on normal and cancerous intestinal epithelial cells. *Gastroenterology*, *139*(3), 918–928.e6. https://doi.org/10.1053/j.gastro.2010.05.081.

Kim, T.-H., Escudero, S., & Shivdasani, R. A. (2012). Intact function of Lgr5 receptor-expressing intestinal stem cells in the absence of Paneth cells. *Proceedings of the National Academy of Sciences of the United States of America*, *109*(10), 3932–3937. https://doi.org/10.1073/pnas.1113890109.

Kim, K.-A., Kakitani, M., Zhao, J., Oshima, T., Tang, T., Binnerts, M., et al. (2005). Mitogenic influence of human R-spondin1 on the intestinal epithelium. *Science*, *309*(5738), 1256–1259. https://doi.org/10.1126/science.1112521.

Kim, T. H., & Shivdasani, R. A. (2011). Genetic evidence that intestinal Notch functions vary regionally and operate through a common mechanism of math1 repression. *Journal of Biological Chemistry*, *286*(13), 11427–11433. https://doi.org/10.1074/jbc.M110.188797.

Kirchberger, S., Royston, D. J., Boulard, O., Thornton, E., Franchini, F., Szabady, R. L., et al. (2013). Innate lymphoid cells sustain colon cancer through production of interleukin-22 in a mouse model. *The Journal of Experimental Medicine, 210*(5), 917–931. https://doi.org/10.1084/jem.20122308.

Komano, H., Fujiura, Y., Kawaguchi, M., Matsumoto, S., Hashimoto, Y., Obana, S., et al. (1995). Homeostatic regulation of intestinal epithelia by intraepithelial gamma delta T cells. *Proceedings of the National Academy of Sciences of the United States of America, 92*(13), 6147–6151. https://doi.org/10.1073/pnas.92.13.6147.

Koo, B. K., Spit, M., Jordens, I., Low, T. Y., Stange, D. E., van de Wetering, M., et al. (2012). Tumour suppressor RNF43 is a stem-cell E3 ligase that induces endocytosis of Wnt receptors. *Nature, 488*(7413), 665–669. https://doi.org/10.1038/nature11308.

Korinek, V., Barker, N., Moerer, P., Van Donselaar, E., Huls, G., Peters, P. J., et al. (1998). Depletion of epithelial stem-cell compartments in the small intestine of mice lacking Tcf-4. *Nature, 19*, 379–383.

Kosinski, C., Li, V. S. W., Chan, A. S. Y., Zhang, J., Ho, C., Tsui, W. Y., et al. (2007). Gene expression patterns of human colon tops and basal crypts and BMP antagonists as intestinal stem cell niche factors. *Proceedings of the National Academy of Sciences of the United States of America, 104*(39), 15418–15423. https://doi.org/10.1073/pnas.0707210104.

Kostic, A. D., Chun, E., Robertson, L., Glickman, J. N., Gallini, C. A., Michaud, M., et al. (2013). Fusobacterium nucleatum potentiates intestinal tumorigenesis and modulates the tumor-immune microenvironment. *Cell Host & Microbe, 14*(2), 207–215. https://doi.org/10.1016/j.chom.2013.07.007.

Kostic, A. D., Gevers, D., Pedamallu, C. S., Kostic, A. D., Gevers, D., Pedamallu, C. S., et al. (2012). Genomic analysis identifies association of Fusobacterium with colorectal carcinoma. *Genome Research, 22*, 292–298. https://doi.org/10.1101/gr.126573.111.

Kraehenbuhl, J.-P., & Neutra, M. R. (2000). Epithelial M cells: Differentiation and function. *Annual Review of Cell and Developmental Biology, 16*(1), 301–332. Retrieved from http://www.annualreviews.org/doi/abs/10.1146/annurev.cellbio.16.1.301%5Cnpapers2://publication/doi/10.1146/annurev.cellbio.16.1.301.

Kuhnert, F., Davis, C. R., Wang, H.-T., Chu, P., Lee, M., Yuan, J., et al. (2004). Essential requirement for Wnt signaling in proliferation of adult small intestine and colon revealed by adenoviral expression of Dickkopf-1. *Proceedings of the National Academy of Sciences of the United States of America, 101*(1), 266–271. https://doi.org/10.1073/pnas.2536800100.

Li, N., Yousefi, M., Nakauka-Ddamba, A., Li, F., Vandivier, L., Parada, K., et al. (2015). The Msi family of RNA-binding proteins function redundantly as intestinal oncoproteins. *Cell Reports, 13*(11), 2440–2455. https://doi.org/10.1016/j.celrep.2015.11.022.

Liaw, D., Marsh, D. J., Li, J., Dahia, P. L., Wang, S. I., Zheng, Z., et al. (1997). Germline mutations of the PTEN gene in Cowden disease, an inherited breast and thyroid cancer syndrome. *Nature Genetics, 16*(1), 64–67. https://doi.org/10.1038/ng0597-64.

Lindemans, C. A., Calafiore, M., Mertelsmann, A. M., O'Connor, M. H., Dudakov, J. A., Jenq, R. R., et al. (2015). Interleukin-22 promotes intestinal-stem-cell-mediated epithelial regeneration. *Nature, 528*(7583), 560–564. https://doi.org/10.1038/nature16460.

Logan, C. Y., & Nusse, R. (2004). The Wnt signaling pathway in development and disease. *Annual Review of Cell and Developmental Biology, 20*(1), 781–810. https://doi.org/10.1146/annurev.cellbio.20.010403.113126.

Lussier, C., Basora, N., Bouatrouss, Y., & Beaulieu, J. F. (2000). Integrins as mediators of epithelial cell–matrix interactions in the human small intestinal mucosa. *Microscopy Research and Technique, 51*(2), 169–178. https://doi.org/10.1002/1097-0029(20001015)51:2<169::AID-JEMT8>3.0.CO;2-A.

Machida-Montani, A., Sasazuki, S., Inoue, M., Natsukawa, S., Shaura, K., Koizumi, Y., et al. (2007). Atrophic gastritis, Helicobacter pylori, and colorectal cancer risk: A case–control study. *Helicobacter, 12*(4), 328–332. https://doi.org/10.1111/j.1523-5378.2007.00513.x.

Madara, J. L. (1982). Cup cells: Structure and distribution of a unique class of epithelial cells in guinea pig, rabbit, and monkey small intestine. *Gastroenterology*, *83*(5), 981–994. https://doi.org/10.1016/s0016-5085(82)80064-4.

Madison, B. B., Braunstein, K., Kuizon, E., Portman, K., Qiao, X. T., & Gumucio, D. L. (2005). Epithelial hedgehog signals pattern the intestinal crypt-villus axis. *Development (Cambridge, England)*, *132*(2), 279–289. https://doi.org/10.1242/dev.01576.

Maher, M. T., Flozak, A. S., Stocker, A. M., Chenn, A., & Gottardi, C. J. (2009). Activity of the β-catenin phosphodestruction complex at cell–cell contacts is enhanced by cadherin-based adhesion. *Journal of Cell Biology*, *186*(2), 219–228. https://doi.org/10.1083/jcb.200811108.

May, R., Riehl, T. E., Hunt, C., Sureban, S. M., Anant, S., & Houchen, C. W. (2008). Identification of a novel putative gastrointestinal stem cell and adenoma stem cell marker, doublecortin and CaM kinase-like-1, following radiation injury and in adenomatous polyposis coli/multiple intestinal neoplasia mice. *Stem Cells*, *26*(3), 630–637. https://doi.org/10.1634/stemcells.2007-0621.

May, R., Sureban, S. M., Hoang, N., Riehl, T. E., Lightfoot, S. A., Ramanujam, R., et al. (2009). Doublecortin and CaM kinase-like-1 and leucine-rich-repeat-containing G-protein-coupled receptor mark quiescent and cycling intestinal stem cells, respectively. *Stem Cells*, *27*(10), 2571–2579. https://doi.org/10.1002/stem.193.

Middendorp, S., Schneeberger, K., Wiegerinck, C. L., Mokry, M., Akkerman, R. D. L., Van Wijngaarden, S., et al. (2014). Adult stem cells in the small intestine are intrinsically programmed with their location-specific function. *Stem Cells*, *32*(5), 1083–1091. https://doi.org/10.1002/stem.1655.

Milano, J., McKay, J., Dagenais, C., Foster-Brown, L., Pognan, F., Gadient, R., et al. (2004). Modulation of Notch processing by gamma-secretase inhibitors causes intestinal goblet cell metaplasia and induction of genes known to specify gut secretory lineage differentiation. *Toxicological Sciences*, *82*(1), 341–358. https://doi.org/10.1093/toxsci/kfh254.

Miyoshi, Y., Ando, H., Nagase, H., Nishisho, I., Horii, a., Miki, Y., et al. (1992). Germ-line mutations of the APC gene in 53 familial adenomatous polyposis patients. *Proceedings of the National Academy of Sciences of the United States of America*, *89*(10), 4452–4456. https://doi.org/10.1073/pnas.89.10.4452.

Montgomery, R. K., Carlone, D. L., Richmond, C. A., Farilla, L., Kranendonk, M. E. G., Henderson, D. E., et al. (2011). Mouse telomerase reverse transcriptase (mTert) expression marks slowly cycling intestinal stem cells. *Proceedings of the National Academy of Sciences of the United States of America*, *108*(1), 179–184. https://doi.org/10.1073/pnas.1013004108.

Monticelli, L. A., Osborne, L. C., Noti, M., Tran, S. V., Zaiss, D. M. W., & Artis, D. (2015). IL-33 promotes an innate immune pathway of intestinal tissue protection dependent on amphiregulin–EGFR interactions. *Proceedings of the National Academy of Sciences of the United States of America*, *112*(34), 10762–10767. https://doi.org/10.1073/pnas.1509070112.

Morin, P. J., Sparks, A. B., Korinek, V., Barker, N., Clevers, H., Vogelstein, B., et al. (1997). Activation of b-catenin–Tcf signaling in colon cancer by mutations in b-catenin or APC. *Science*, *275*, 1787–1790.

Muncan, V., Sansom, O. J., Tertoolen, L., Phesse, T. J., Begthel, H., Sancho, E., et al. (2006). Rapid loss of intestinal crypts upon conditional deletion of the Wnt/Tcf-4 target gene c-Myc. *Molecular and Cellular Biology*, *26*(22), 8418–8426. https://doi.org/10.1128/MCB.00821-06.

Muñoz, J., Stange, D. E., Schepers, A. G., van de Wetering, M., Koo, B.-K., Itzkovitz, S., et al. (2012). The Lgr5 intestinal stem cell signature: Robust expression of proposed quiescent "+4" cell markers. *The EMBO Journal*, *31*(14), 3079–3091. https://doi.org/10.1038/emboj.2012.166.

Murgia, C., Blaikie, P., Kim, N., Dans, M., Petrie, H. T., & Giancotti, F. G. (1998). Cell cycle and adhesion defects in mice carrying a targeted deletion of the integrin β4 cytoplasmic domain. *The EMBO Journal, 17*(14), 3940–3951.

Myant, K. B., Cammareri, P., McGhee, E. J., Ridgway, R. A., Huels, D. J., Cordero, J. B., et al. (2013). ROS production and NF-kB activation triggered by RAC1 facilitate WNT-driven intestinal stem cell proliferation and colorectal cancer initiation. *Cell Stem Cell, 12*(6), 761–773. https://doi.org/10.1016/j.stem.2013.04.006.

Neal, M. D., Sodhi, C. P., Jia, H., Dyer, M., Egan, C. E., Yazji, I., et al. (2012). Toll-like receptor 4 is expressed on intestinal stem cells and regulates their proliferation and apoptosis via the p53 up-regulated modulator of apoptosis. *Journal of Biological Chemistry, 287*(44), 37296–37308. https://doi.org/10.1074/jbc.M112.375881.

Nigro, G., Rossi, R., Commere, P. H., Jay, P., & Sansonetti, P. J. (2014). The cytosolic bacterial peptidoglycan sensor Nod2 affords stem cell protection and links microbes to gut epithelial regeneration. *Cell Host & Microbe, 15*(6), 792–798. https://doi.org/10.1016/j.chom.2014.05.003.

Obata, Y., Takahashi, D., Ebisawa, M., Kakiguchi, K., Yonemura, S., Jinnohara, T., et al. (2012). Epithelial cell-intrinsic notch signaling plays an essential role in the maintenance of gut immune homeostasis. *The Journal of Immunology, 188*(5), 2427–2436. https://doi.org/10.4049/jimmunol.1101128.

Pan, M. R., Peng, G., Hungs, W. C., & Lin, S. Y. (2011). Monoubiquitination of H2AX protein regulates DNA damage response signaling. *Journal of Biological Chemistry, 286*(32), 28599–28607. https://doi.org/10.1074/jbc.M111.256297.

Pédron, T., Mulet, C., Dauga, C., Frangeul, L., Chervaux, C., Grompone, G., et al. (2012). A crypt-specific core microbiota resides in the mouse colon. *mBio, 3*(3), e00116-12. https://doi.org/10.1128/mBio.00116-12.

Peignon, G., Durand, A., Cacheux, W., Ayrault, O., Terris, B., Laurent-Puig, P., et al. (2011). Complex interplay between b-catenin signalling and Notch effectors in intestinal tumorigenesis. *Gut, 60*, 166–176. https://doi.org/10.1136/gut.2009.204719.

Pellegrinet, L., Rodilla, V., Liu, Z., Chen, S., Koch, U., Espinosa, L., et al. (2011). Dll1- and Dll4-mediated notch signaling are required for homeostasis of intestinal stem cells. *Gastroenterology, 140*(4), 1230–1240. https://doi.org/10.1053/j.gastro.2011.01.005.

Pinto, D., Gregorieff, A., Begthel, H., & Clevers, H. (2003). Canonical Wnt signals are essential for homeostasis of the intestinal epithelium. *Genes and Development, 17*(14), 1709–1713. https://doi.org/10.1101/gad.267103.

Potten, C. S., Booth, C., & Pritchard, D. M. (1997). The intestinal epithelial stem cell: The mucosal governor. *International Journal of Experimental Pathology, 78*(4), 219–243. https://doi.org/10.1046/j.1365-2613.1997.280362.x.

Potten, C. S., Booth, C., Tudor, G. L., Booth, D., Brady, G., Hurley, P., et al. (2003). Identification of a putative intestinal stem cell and early lineage marker; musashi-1. *Differentiation, 71*(1), 28–41. https://doi.org/10.1046/j.1432-0436.2003.700603.x.

Potten, C. S., Hume, W. J., Reid, P., & Cairns, J. (1978). The segregation of DNA in epithelial stem cells. *Cell, 15*(3), 899–906. https://doi.org/10.1016/0092-8674(78)90274-X.

Potten, C. S., & Loeffler, M. (1987). A comprehensive model of the crypts of the small intestine of the mouse provides insight into the mechanisms of cell migration and the proliferation hierarchy. *Journal of Theoretical Biology, 127*(4), 381–390.

Powell, D. W., Pinchuk, I. V., Saada, J. I., Chen, X., & Mifflin, R. C. (2011). Mesenchymal cells of the intestinal lamina propria. *Annual Review of Physiology, 73*(1), 213–237. https://doi.org/10.1146/annurev.physiol.70.113006.100646.

Powell, A. E., Wang, Y., Li, Y., Poulin, E. J., Means, A. L., Washington, M. K., et al. (2012). The pan-ErbB negative regulator lrig1 is an intestinal stem cell marker that functions as a tumor suppressor. *Cell, 149*(1), 146–158. https://doi.org/10.1016/j.cell.2012.02.042.

Powell, S. M., Zilz, N., Beazer-Barclay, Y., Bryan, T. M., Hamilton, S. R., Thibodeau, S. N., et al. (1992). APC mutations occur early during colorectal tumorigenesis. *Nature, 359*(6392), 235–237. https://doi.org/10.1038/359235a0.

Pütsep, K., Axelsson, L. G., Boman, A., Midtvedt, T., Normark, S., Boman, H. G., et al. (2000). Germ-free and colonized mice generate the same products from enteric prodefensins. *Journal of Biological Chemistry, 275*(51), 40478–40482. https://doi.org/10.1074/jbc.M007816200.

Qi, Z., Li, Y., Zhao, B., Xu, C., Liu, Y., Li, H., et al. (2017). BMP restricts stemness of intestinal Lgr5+ stem cells by directly suppressing their signature genes. *Nature Communications, 8*, 13824. https://doi.org/10.1038/ncomms13824.

Rakoff-Nahoum, S., & Medzhitov, R. (2007). Regulation of spontaneous intestinal tumorigenesis through the adaptor protein MyD88. *Science, 317*(5834), 124–127. https://doi.org/10.1126/science.1140488.

Rakoff-Nahoum, S., Pglino, J., Eslami-Varzaneh, F., Edberg, S., & Medzhitov, R. (2004). Recognition of commensal microflora by toll-like receptors in required for intestinal homeostasis. *Cell, 118*, 229–241.

Reed, K. R., Tunster, S. J., Young, M., Carrico, A., John, R. M., & Clarke, A. R. (2012). Entopic overexpression of *Ascl2* does not accelerate tumourigenesis in Apc^{Min} mice. *Gut, 61*(10), 1435–1438. https://doi.org/10.1136/gutjnl-2011-300842.

Reikvam, D. H., Erofeev, A., Sandvik, A., Grcic, V., Jahnsen, F. L., Gaustad, P., et al. (2011). Depletion of murine intestinal microbiota: Effects on gut mucosa and epithelial gene expression. *PLoS One, 6*(3), 1–13. https://doi.org/10.1371/journal.pone.0017996.

Riccio, O., van Gijn, M. E., Bezdek, A. C., Pellegrinet, L., van Es, J. H., Zimber-Strobl, U., et al. (2008). Loss of intestinal crypt progenitor cells owing to inactivation of both Notch1 and Notch2 is accompanied by derepression of CDK inhibitors p27Kip1 and p57Kip2. *EMBO Reports, 9*(4), 377–383. https://doi.org/10.1038/embor.2008.7.

Ritsma, L., Ellenbroek, S. I. J., Zomer, A., Snippert, H. J., De Sauvage, F. J., Simons, B. D., et al. (2014). Intestinal crypt homeostasis revealed at single-stem-cell level by in vivo live imaging. *Nature, 507*, 362–365. https://doi.org/10.1038/nature12972.

Rodríguez-Colman, M. J., Schewe, M., Meerlo, M., Stigter, E., Gerrits, J., Pras-Raves, M., et al. (2017). Interplay between metabolic identities in the intestinal crypt supports stem cell function. *Nature, 543*, 424–427. https://doi.org/10.1038/nature21673.

Romagnolo, B., Berrebi, D., Saadi-Keddoucci, S., Porteu, A., Pichard, A. L., Peuchmaur, M., et al. (1999). Intestinal dysplasia and adenoma in transgenic mice after overexpression of an activated β-catenin. *Cancer Research, 59*, 3875–3879.

Rothenberg, M. E., Nusse, Y., Kalisky, T., Lee, J. J., Dalerba, P., Scheeren, F., et al. (2012). Identification of a cKit+ colonic crypt base secretory cell that supports Lgr5+ stem cells in mice. *Gastroenterology, 142*(5), 1195–1205.e6. https://doi.org/10.1053/j.gastro.2012.02.006.

Rubin, D. C., Roth, K. a., Birkenmeier, E. H., & Gordon, J. I. (1991). Epithelial cell differentiation in normal and transgenic mouse intestinal isografts. *The Journal of Cell Biology, 113*(5), 1183–1192.

Rubin, D. C., Swietlicki, E., Roth, K. A., & Gordon, J. I. (1992). Use of fetal intestinal isografts from normal and transgenic mice to study the programming of positional information along the duodenal-to-colonic axis. *Journal of Biological Chemistry, 267*(21), 15122–15133.

Salzman, N. H., Hung, K., Haribhai, D., Chu, H., Karlsson-Sjöberg, J., Amir, E., et al. (2010). Enteric defensins are essential regulators of intestinal microbial ecology. *Nature Immunology, 11*(1), 76–83. https://doi.org/10.1038/ni.1825.

San Roman, A. K., Jayewickreme, C. D., Murtaugh, L. C., & Shivdasani, R. A. (2014). Wnt secretion from epithelial cells and subepithelial myofibroblasts is not required in the mouse intestinal stem cell niche in vivo. *Stem Cell Reports, 2*(2), 127–134. https://doi.org/10.1016/j.stemcr.2013.12.012.

Sander, G. R., & Powell, B. C. (2004). Expression of notch receptors and ligands in the adult gut. *The Journal of Histochemistry and Cytochemistry: Official Journal of the Histochemistry Society, 52*(4), 509–516. https://doi.org/10.1177/002215540405200409.

Sangiorgi, E., & Capecchi, M. R. (2008). Bmi1 is expressed in vivo in intestinal stem cells. *Nature Genetics, 40*(7), 915–920. https://doi.org/10.1038/ng.165.

Sansom, O. J., Reed, K. R., Hayes, A. J., Ireland, H., Brinkmann, H., Newton, I. P., et al. (2004). Loss of Apc in vivo immediately perturbs Wnt signaling, differentiation, and migration. *Genes and Development, 18*(12), 1385–1390. https://doi.org/10.1101/gad.287404.

Sasaki, N., Sachs, N., Wiebrands, K., Ellenbroek, S. I., Fumagalli, A., Lyubimova, A., et al. (2016). Reg4 + deep crypt secretory cells function as epithelial niche for Lgr5 + stem cells in colon. *Proceedings of the National Academy of Sciences of the United States of America, 113*, E5399–E5407.

Sato, T., Stange, D. E., Ferrante, M., Vries, R. G. J., Van Es, J. H., Van Den Brink, S., et al. (2011). Long-term expansion of epithelial organoids from human colon, adenoma, adenocarcinoma, and Barrett's epithelium. *Gastroenterology, 141*(5), 1762–1772. https://doi.org/10.1053/j.gastro.2011.07.050.

Sato, T., van Es, J. H., Snippert, H. J., Stange, D. E., Vries, R. G., van den Born, M., et al. (2011). Paneth cells constitute the niche for Lgr5 stem cells in intestinal crypts. *Nature, 469*(7330), 415–418. https://doi.org/10.1038/nature09637.

Sato, T., Vries, R. G., Snippert, H. J., van de Wetering, M., Barker, N., Stange, D. E., et al. (2009). Single Lgr5 stem cells build crypt-villus structures in vitro without a mesenchymal niche. *Nature, 459*(7244), 262–265. https://doi.org/10.1038/nature07935.

Savage, D. (1977). Microbial ecology of the gastrointestinal tract. *Annual Review of Microbiology, 31*, 107–133.

Schepers, A. G., Snippert, H. J., Stange, D. E., van den Born, M., van Es, J. H., van de Wetering, M., et al. (2012). Lineage tracing reveals Lgr5 + stem cell activity in mouse intestinal adenomas. *Science, 337*, 819–824. https://doi.org/10.1126/science.1229223.

Schepers, A. G., Vries, R., van den Born, M., van de Wetering, M., & Clevers, H. (2011). Lgr5 intestinal stem cells have high telomerase activity and randomly segregate their chromosomes. *The EMBO Journal, 30*(6), 1104–1109. https://doi.org/10.1038/emboj.2011.26.

Schmidt, G. H., Wilkinson, M. M., & Ponder, B. A. J. (1985). Cell migration pathway in the intestinal epithelium: An in situ marker system using mouse aggregation chimeras. *Cell, 40*(2), 425–429. https://doi.org/10.1016/0092-8674(85)90156-4.

Schneider, M. R., Dahlhoff, M., Horst, D., Hirschi, B., Trülzsch, K., Müller-Höcker, J., et al. (2010). A key role for E-cadherin in intestinal homeostasis and Paneth cell maturation. *PLoS One, 5*(12), e14325. https://doi.org/10.1371/journal.pone.0014325.

Schröder, N., & Gossler, A. (2002). Expression of Notch pathway components in fetal and adult mouse small intestine. *Gene Expression Patterns, 2*, 247–250. https://doi.org/10.1016/S1.

Schuijers, J., Junker, J. P., Mokry, M., Hatzis, P., Koo, B. K., Sasselli, V., et al. (2015). Ascl2 acts as an R-spondin/Wnt-responsive switch to control stemness in intestinal crypts. *Cell Stem Cell, 16*, 158–170. https://doi.org/10.1016/j.stem.2014.12.006.

Schuijers, J., Van Der Flier, L. G., Van Es, J., & Clevers, H. (2014). Robust Cre-mediated recombination in small intestinal stem cells utilizing the Olfm4 locus. *Stem Cell Reports, 3*(2), 234–241. https://doi.org/10.1016/j.stemcr.2014.05.018.

Seshagiri, S., Stawiski, E. W., Durinck, S., Modrusan, Z., Storm, E. E., Conboy, C. B., et al. (2012). Recurrent R-spondin fusions in colon cancer. *Nature, 488*(7413), 660–664. https://doi.org/10.1038/nature11282.

Shroyer, N. F., Helmrath, M. A., Wang, V. Y. C., Antalffy, B., Henning, S. J., & Zoghbi, H. Y. (2007). Intestine-specific ablation of mouse atonal homolog 1 (Math1) reveals a role in cellular homeostasis. *Gastroenterology, 132*(7), 2478–2488. https://doi.org/10.1053/j.gastro.2007.03.047.

Sneddon, J. B., Zhen, H. H., Montgomery, K., van de Rijn, M., Tward, A. D., West, R., et al. (2006). Bone morphogenetic protein antagonist gremlin 1 is widely expressed by cancer-associated stromal cells and can promote tumor cell proliferation. *Proceedings of the National Academy of Sciences of the United States of America, 103*(40), 14842–14847. https://doi.org/10.1073/pnas.0606857103.

Snippert, H. J., van der Flier, L. G., Sato, T., van Es, J. H., van den Born, M., Kroon-Veenboer, C., et al. (2010). Intestinal crypt homeostasis results from neutral competition between symmetrically dividing Lgr5 stem cells. *Cell, 143*(1), 134–144. https://doi.org/10.1016/j.cell.2010.09.016.

Solanas, G., Cortina, C., Sevillano, M., & Batlle, E. (2011). Cleavage of E-cadherin by ADAM10 mediates epithelial cell sorting downstream of EphB signalling. *Nature Cell Biology, 13*(9), 1100–1107. https://doi.org/10.1038/ncb2298.

Stanger, B. Z., Datar, R., Murtaugh, L. C., & Melton, D. A. (2005). Direct regulation of intestinal fate by Notch. *Proceedings of the National Academy of Sciences of the United States of America, 102*(35), 12443–12448. https://doi.org/10.1073/pnas.0505690102.

Stzepourginski, I., Nigro, G., Jacob, J., Dulauroy, S., Sansonetti, P. J., et al. (2016). CD34+ mesenchymal cells are a major component of the intestinal stem cells niche at homeostasis and after injury. *Proceedings of the National Academy of Sciences of the United States of America, 114*(4), E506–E513.

Takeda, N., Jain, R., LeBoeuf, M. R., Wang, Q., Lu, M. M., & Epstein, J. A. (2011). Interconversion between intestinal stem cell populations in distinct niches. *Science, 334*(6061), 1420–1424. https://doi.org/10.1126/science.1213214.

Taketo, M. M., & Edelmann, W. (2009). Mouse models of colon cancer. *Gastroenterology, 136*(3), 780–798.

Tetteh, P. W., Basak, O., Farin, H. F., Wiebrands, K., Kretzschmar, K., Begthel, H., et al. (2016). Replacement of lost lgr5-positive stem cells through plasticity of their enterocyte-lineage daughters. *Cell Stem Cell, 31*(14). https://doi.org/10.1016/j.stem.2016.01.001.

Tian, H., Biehs, B., Warming, S., Leong, K. G., Rangell, L., Klein, O. D., et al. (2011). A reserve stem cell population in small intestine renders Lgr5-positive cells dispensable. *Nature, 478*(7368), 255–259. https://doi.org/10.1038/nature10408.

Tsai, Y. H., Vandussen, K. L., Sawey, E. T., Wade, A. W., Kasper, C., Rakshit, S., et al. (2014). ADAM10 regulates notch function in intestinal stem cells of mice. *Gastroenterology, 147*(4), 822–834.e13. https://doi.org/10.1053/j.gastro.2014.07.003.

Ueo, T., Imayoshi, I., Kobayashi, T., Ohtsuka, T., Seno, H., Nakase, H., et al. (2012). The role of Hes genes in intestinal development, homeostasis and tumor formation. *Development, 139*(6), 1071–1082.

Vaishnava, S., Behrendt, C. L., Ismail, A. S., Eckmann, L., & Hooper, L. V. (2008). Paneth cells directly sense gut commensals and maintain homeostasis at the intestinal host–microbial interface. *Proceedings of the National Academy of Sciences of the United States of America, 105*(52), 20858–20863. https://doi.org/10.1073/pnas.0808723105.

Valenta, T., Degirmenci, B., Moor, A. E., Herr, P., Zimmerli, D., Moor, M. B., et al. (2016). Wnt ligands secreted by subepithelial mesenchymal cells are essential for the survival of intestinal stem cells and gut homeostasis. *Cell Reports, 15*, 911–918. https://doi.org/10.1016/j.celrep.2016.03.088.

van den Brink, G. R., Bleuming, S. a., Hardwick, J. C. H., Schepman, B. L., Offerhaus, G. J., Keller, J. J., et al. (2004). Indian Hedgehog is an antagonist of Wnt signaling in colonic epithelial cell differentiation. *Nature Genetics, 36*(3), 277–282. https://doi.org/10.1038/ng1304.

van der Flier, L. G., Haegebarth, A., Stange, D. E., van de Wetering, M., & Clevers, H. (2009). OLFM4 is a robust marker for stem cells in human intestine and marks a subset of colorectal cancer cells. *Gastroenterology, 137*(1), 15–17. https://doi.org/10.1053/j.gastro.2009.05.035.

van der Flier, L. G., van Gijn, M. E., Hatzis, P., Kujala, P., Haegebarth, A., Stange, D. E., et al. (2009). Transcription factor achaete scute-like 2 controls intestinal stem cell fate. *Cell, 136*(5), 903–912. https://doi.org/10.1016/j.cell.2009.01.031.

van Dop, W. A., Uhmann, A., Wijgerde, M., Sleddens-Linkels, E., Heijmans, J., Offerhaus, G. J., et al. (2009). Depletion of the colonic epithelial precursor cell compartment upon conditional activation of the hedgehog pathway. *Gastroenterology, 136*(7), 2195–2203.e7. https://doi.org/10.1053/j.gastro.2009.02.068.

van Es, J. H., de Geest, N., van de Born, M., Clevers, H., & Hassan, B. A. (2010). Intestinal stem cells lacking the Math1 tumour suppressor are refractory to Notch inhibitors. *Nature Communications, 1*(2), 1–5. https://doi.org/10.1038/ncomms1017.

van Es, J. H., Haegebarth, A., Kujala, P., Itzkovitz, S., Koo, B.-K., Boj, S. F., et al. (2012). A critical role for the Wnt effector Tcf4 in adult intestinal homeostatic self-renewal. *Molecular and Cellular Biology, 32*(10), 1918–1927. https://doi.org/10.1128/MCB.06288-11.

van Es, J. H., Jay, P., Gregorieff, A., van Gijn, M. E., Jonkheer, S., Hatzis, P., et al. (2005). Wnt signalling induces maturation of Paneth cells in intestinal crypts. *Nature Cell Biology*, 7(4), 381–386. https://doi.org/10.1038/ncb1240.

van Es, J. H., Sato, T., van de Wetering, M., Lyubimova, A., Yee Nee, A. N., Gregorieff, A., et al. (2012). Dll1+ secretory progenitor cells revert to stem cells upon crypt damage. *Nature Cell Biology, 14*(10), 1099–1104. https://doi.org/10.1038/ncb2581.

van Es, J. H., van Gijn, M. E., Riccio, O., van den Born, M., Vooijs, M., Begthel, H., et al. (2005). Notch/γ-secretase inhibition turns proliferative cells in intestinal crypts and adenomas into goblet cells. *Nature, 435*(7044), 959–963. https://doi.org/10.1038/nature03659.

VanDussen, K. L., Carulli, a. J., Keeley, T. M., Patel, S. R., Puthoff, B. J., Magness, S. T., et al. (2012). Notch signaling modulates proliferation and differentiation of intestinal crypt base columnar stem cells. *Development, 139*(3), 488–497. https://doi.org/10.1242/dev.070763.

Waite, K. A., Waite, K. A., & Eng, C. (2003). BMP2 exposure results in decreased PTEN protein degradation and increased PTEN levels BMP2 exposure results in decreased PTEN protein degradation and increased PTEN levels. *Human Molecular Genetics, 12*(6), 679–684. https://doi.org/10.1093/hmg/ddg069.

Wong, M. H., Hermiston, M. L., Syder, A. J., Gordon, J. I., & Kipnis, D. M. (1996). Forced expression of the tumor suppressor adenomatosis polyposis coli protein induces disordered cell migration in the intestinal epithelium (catenins/E-cadherin/embryonic stem cells/chimeric mice). *Cell Biology, 93*, 9588–9593.

Wong, V. W. Y., Stange, D. E., Page, M. E., Buczacki, S., Wabik, A., Itami, S., et al. (2012). Lrig1 controls intestinal stem-cell homeostasis by negative regulation of ErbB signalling. *Nature Cell Biology, 14*(4), 401–408. https://doi.org/10.1038/ncb2464.

Worthley, D. L., Giraud, A. S., & Wang, T. C. (2010). The extracellular matrix in digestive cancer. *Cancer Microenvironment, 3*(1), 177–185. https://doi.org/10.1007/s12307-010-0053-4.

Wright, N. A., & Irwin, M. (1982). The kinetics of villus cell populations in the mouse small intestine. I. Normal villi: The steady state requirement. *Cell and Tissue Kinetics, 15*(6), 595–609.

Yan, K., Chia, L., & Li, X. (2012). The intestinal stem cell markers Bmi1 and Lgr5 identify two functionally distinct populations. *PNAS, 109*(2), 466–471. https://doi.org/10.1073/pnas.1118857109/-/DCSupplemental.www.pnas.org/cgi/doi/10.1073/pnas.1118857109.

Yan, K. S., Gevaert, O., Zheng, G. X. Y., Anchang, B., Probert, C. S., Larkin, K. A., et al. (2017). Intestinal enteroendocrine lineage cells possess homeostatic and injury-inducible stem cell activity. *Cell Stem Cell, 21*(1), 78–90.

Yang, Q., Bermingham, N. a., Finegold, M. J., & Zoghbi, H. Y. (2001). Requirement of Math1 for secretory cell lineage commitment in the mouse intestine. *Science (New York, NY)*, *294*(5549), 2155–2158. https://doi.org/10.1126/science.1065718.

Yang, Y. P., Ma, H., Starchenko, A., Huh, W. J., Li, W., Hickman, F. E., et al. (2017). A chimeric Egfr protein reporter mouse reveals Egfr localization and trafficking in vivo. *Cell Reports*, *19*(6), 1257–1267. https://doi.org/10.1016/j.celrep.2017.04.048.

Yilmaz, H., Katajisto, P., Lamming, D. W., Gu, Y., Birsoy, K., Dursun, A., et al. (2012). mTORC1 in the Paneth cell niche couples intestinal stem-cell function to calorie intake. *Nature*, *486*, 490–495. https://doi.org/10.1038/nature11163.

Yin, X., Farin, H. F., van Es, J. H., Clevers, H., Langer, R., & Karp, J. M. (2014). Niche-independent high-purity cultures of Lgr5 + intestinal stem cells and their progeny. *Nature Methods*, *11*(1), 106–112.

Zhu, L., Gibson, P., Currle, D. S., Tong, Y., Richardson, R. J., Bayazitov, I. T., et al. (2009). Prominin 1 marks intestinal stem cells that are susceptible to neoplastic transformation. *Nature*, *457*(7229), 603–607. https://doi.org/10.1038/nature07589.

CHAPTER TWO

Multitasking Paneth Cells in the Intestinal Stem Cell Niche

Matthias Schewe[1], Riccardo Fodde[1]
Erasmus MC Cancer Institute, Erasmus University Medical Center, Rotterdam, The Netherlands
[1]Corresponding authors: e-mail address: matthias.schewe@helsinki.fi; r.fodde@erasmusmc.nl

Contents

1. Paneth Cells: From Schwalbe and Paneth to the Discovery of Their Secretory Role 41
   1.1 Early Studies Anticipate Paneth Cells' Multitasking Functional Nature 41
   1.2 PCs: The Secretory Bodyguards of the Crypt 43
2. The Intestinal Crypt Stem Cell Niche 46
   2.1 The Discovery of the Intestinal Stem Cell 46
   2.2 PCs and Crypt Development 48
   2.3 Paneth-Specific Signaling Pathways: Wnt and Notch 50
   2.4 PCs Constitute the Main Epithelial Intestinal Stem Cell Niche 55
   2.5 PCs and Nutrient Sensing: A Matter of Metabolism? 57
   2.6 Culturing Intestinal Crypts ex vivo as Organoids 58
3. PCs, Dysbiosis, and IBD 60
   3.1 Nod2 in IBD and PCs 62
   3.2 ER Stress and Unfolded Protein Response in IBD and PCs 63
   3.3 Autophagy in IBD and PCs 64
   3.4 The Role of Immunity and Environmental Factors 66
   3.5 PCs as IBD-Specific Disease Markers 66
   3.6 PCs as Stem Cell Niche in Inflammation 67
4. Conclusive Remarks 67
   4.1 Multitasking PCs in Health and Disease 67
References 68

1. PANETH CELLS: FROM SCHWALBE AND PANETH TO THE DISCOVERY OF THEIR SECRETORY ROLE

1.1 Early Studies Anticipate Paneth Cells' Multitasking Functional Nature

Granulated cells in the fundus of the crypts of Lieberkühn were first described in 1872, along with the other cells of the intestinal epithelium,

Advances in Stem Cells and their Niches, Volume 2
ISSN 2468-5097
https://doi.org/10.1016/bs.asn.2017.12.001

by Schwalbe, a German anatomist, histologist, and anthropologist (Schwalbe, 1872). Nevertheless, they take their name from Joseph Paneth, an Austrian histologist and physiologist who first performed their detailed morphological analysis in 1888 (Paneth, 1888). For many decades after their discovery, however, the role of Paneth cells (PCs) remained obscure due to mainly observational studies limited by the available methods and mostly focused on the morphological features of PCs and their granular payload.

In 1899 Möller reported on the presence of PCs in several animal species including mouse, guinea pig, hox, sheep, and horse (Möller, 1899). Given the broad spectrum of species found to host PCs and to feature acidophilic granules and mucus in the small intestinal epithelial lining, the question arose whether they represented independent zymogenic cells or were involved in mucus production (Bizzozero, 1892). Bizzozero claimed to have found cells with a dual Paneth/goblet cell fate, therefore raising the possibility of PCs representing goblet cells precursors (Bizzozero, 1892). Few years later, Klein brought evidence supporting the independent zymogenic function of PCs, distinct from goblet cells, based on staining analysis with mucus-specific dyes (Klein, 1906). Mols and Cowdry studied the effects of different diets in murine PCs and observed that their intracellular content changed depending on the specific diet fed (Cowdry, 1934; Georges, 1930), thus suggesting, already at these very early days, specific roles for this cell type in metabolism and nutrient sensing to be confirmed nearly a century later (Rodriguez-Colman et al., 2017; Yilmaz et al., 2012).

In 1937 Hertzog analyzed PCs in man and reported on their reduced abundancy in the duodenum. Also PCs multiplicity and distribution did not appear to be influenced by sex or age and, of note, they occasionally appeared in colon, stomach, and appendix in association with pathological conditions (Hertzog, 1937). The latter observation led to additional studies on the alleged function of PCs in disease (Creamer, 1967a, 1967b; Creamer & Pink, 1967; Lewin, 1969a; Trier, 1966, 1967). The most accurate description came from studies by Lewin who described and counted the appearance and number of PCs in the small and large intestine. Of note, PC multiplicity was shown to be affected upon inflammation depending on the specific type of pathology and the degree of the inflammatory damage. Decreased PC numbers were found in acute inflammation of the small intestine and in Crohn's disease (CD). Interestingly, PCs were also found to proliferate upon inflammation (Lewin, 1969a, 1969b, 1969c). Contrary to what observed in the small intestine, an increased number of PCs was found in the colon in association with ulcerative colitis (UC) and other inflammatory diseases of the distal bowel. As of today, it is still unclear whether Paneth cell metaplasia (PCM), i.e., the appearance of PCs in the large intestine, plays any functional role in the tissue response to inflammatory insults and in the increased colon cancer risk associated with inflammatory bowel disease (IBD). The role of PCs in IBD will be discussed further in depth in a later section of this chapter.

PCs have also been observed within intestinal tumors: several studies have reported on the presence of lysozyme-positive neoplastic cells in adenomas and carcinomas of the colon, often with proliferative features. The number of PCs in colonic tumors was higher in proximal lesions when compared with distal neoplasias (Lewin, 1968). Notably, PCs have been observed in colonic tumors from patients with but also without a history of IBDs. As for the biogenesis of PCM in the colonic mucosa, early studies proposed that these Paneth-like cells arise through metaplastic changes brought about by the altered inflammatory and/or neoplastic milieu (Lewin, 1968). The term Paneth cell *metaplasia* (from the Greek *metaplasmos* = change in form) was therefore employed to indicate the conversion of a cell type into another. Metaplastic PCs have been found in a broad spectrum of pathological conditions ranging from IBDs to cancer (Paterson & Watson, 1961; Symonds, 1974).

The origin and fate of PCs and their context-dependent proliferative state were investigated by Cheng and Leblond who found that PCs exist in a variety of maturation stages in vivo as a measure of the size of their granules (23% with granules smaller than 2 μM and classified as "young" PCs; 63% with granules between 2 and 3 μM; and 14% "old" PCs with granules larger than 3 μM) (Cheng, 1974). These studies also showed that PCs were postmitotic in homeostatic conditions and persisted for long terms at the bottom of the crypt, unlike most intestinal epithelial cells that actively migrate upward along the crypt–villus axis to be then exfoliated by apoptosis within 5–6 days.

1.2 PCs: The Secretory Bodyguards of the Crypt

The large acidophilic granules apically secreted into the crypt lumen that earmark PCs have been the object of several electron microscopy studies. The secretory function of the PC is prominent throughout its different subcellular compartments featuring an extensive Golgi apparatus and very pronounced endoplasmic reticulum. However, the function of the secretory granules remained elusive until lysozyme was identified as one of their main components (Deckx, Vantrappen, & Parein, 1967). As lysozyme is a strong antibacterial agent, a role was proposed for PCs in host defense mechanisms in response to bacterial flora.

Subsequently, two additional families of antimicrobial proteins, namely, secretory phospholipases (e.g., *Pla2g2a*) (Senegas-Balas et al., 1984), endowed with additional lipid modifying function, and α-defensins (also known as cryptdins for their location at the bottom of the crypt) (Ouellette & Cordell, 1988; Ouellette, Miller, Henschen, & Selsted, 1992) were found to be expressed by murine PCs and specifically localized in their secretory granules.

These studies sparked several additional efforts toward the elucidation of the mechanisms underlying PC secretion and the environmental stimuli that trigger it. To this aim, whole crypts were detached from the lamina propria and exposed to bacteria such as *Salmonella typhimurium*, *Escherichia coli*, or *Staphylococcus aureus*. A potent induction of granule secretion and consequent antibacterial activity was observed upon stimulation of these ex vivo isolated crypts with bacterial cells. More than 90% of the microorganisms to which the crypts were exposed were killed. Notably, most of the antimicrobial activity appeared to be elicited by cryptdins, as shown by the observed 70% reduction in bactericidal function of the granules upon exposure to antibodies directed against cryptdin and cryptdin-like enzymes. Exocytosis of PC granules was also observed upon exposure to bacterial surface products such as lipopolysaccharide (LPS) in a dose-dependent manner (Ayabe et al., 2000). Furthermore, granule secretion can be triggered by a broad range of stimuli including acetylcholinergic (Qu, Lloyd, Walsh, & Lehrer, 1996; Satoh, 1988) and toll-like receptor agonists (Rumio et al., 2004, 2012). More recent studies suggest that secretion of the granules by PCs is triggered by bacterial cells or antigens in an indirect way and requires interferon gamma (IFN-γ), mainly secreted by leukocytes coisolated with the crypts. To prove that stimulation of PC secretion is indirectly stimulated by bacterial antigens, crypts were cultured ex vivo as self-renewing organoids or "miniguts" (this technique is more accurately described further in this chapter) and the basal or luminal side of such structures exposed to LPS, flagellin, peptidoglycan (PGN), heat-inactivated *E. coli*, and live bacterial cells (Farin et al., 2014). No change in PC morphology or granule secretion was observed after 12 h. Furthermore, no changes in intracellular or secreted lysozyme expression were observed. PC expression of bacterial receptors like Cd14 and Toll-like receptors 2,3,7,8, and 9 remained unchanged in the cultured organoids when compared to PCs in vivo, the latter also indicating that these receptors are retained in ex vivo cultured crypts.

Consequently, several inflammatory molecules were tested for their capacity to induce discharge of PC granules. Among several molecules including interleukin 22 (IL22) and 6 (IL6), tumor necrosis factor, and IFN-γ, only the latter proved to be a potent inducer of granule secretion by PCs (Farin et al., 2014).

Additional in vivo experimental evidence for the role of IFN-γ as the main inflammatory molecule responsible for granule extrusion was provided by injecting wild-type and IFN-γ knockout mice with anti-CD3 antibody to trigger T cell activation and IFN-γ expression. When wild-type mice were administered the CD3 antibody, PCs degranulated as shown by decreased lysozyme content. Notably, no PC degranulation or any decrease in PC multiplicity was observed in in CD3-treated IFN-γ knockout mice (Farin et al., 2014).

PC granule secretion was also shown to be mediated by calcium fluxes. By tracking Ca^{2+} dynamics in ex vivo crypts during granules secretion, a biphasic increase in cytosolic Ca^{2+} was revealed with the first phase mainly arising from intracellular storage, whereas the second depends on extracellular calcium uptake (Satoh, Habara, Ono, & Kanno, 1995). The second rise in cytosolic Ca^{2+} can be reduced by selectively blocking the calcium-activated potassium channel *IKCa1* (by means of clotrimazole or charybdotoxin), and by measuring the levels of cryptdin secretion. The observed 50% reduction in secretion also corresponded to a similar reduction in antibacterial activity of crypts exposed to *S. typhimurium* (Ayabe et al., 2000, 2002; Satoh et al., 1995).

The antimicrobial peptides (AMPs) secreted by PCs have been proposed to encompass two main functions: (i) protect the host from enteric pathogens and (ii) keep in balance and shape the composition of the microbiome (Vaishnava, Behrendt, Ismail, Eckmann, & Hooper, 2008). These two deeply intertwined functions are of great relevance since the composition and number of PCs might unfavorably alter the composition of microbiota, a condition known as dysbiosis, thus rendering the host more susceptible to a variety of infectious and noninfectious diseases (Candela et al., 2011; Dumas et al., 2006; Karlsson et al., 2012; Koren et al., 2011; Salzman & Bevins, 2013; Turnbaugh et al., 2006).

The analysis and study of AMPs like defensins in the mouse are made more difficult by the complex genomic structure encompassing several gene paralogs within a single locus, and by the strain-dependent composition of AMPs among inbred mice (Amid et al., 2009; Gulati et al., 2012; Shanahan, Tanabe, & Ouellette, 2011).

As previously stated, PCs are found in a wide range of species (Möller, 1899). Likewise, AMPs have also been conserved throughout evolution (Ouellette, 2011). One notable difference is represented by the presence of multiple defensins expressed by PCs in a broad spectrum of mammalian species, whereas their human counterparts only encode for two, i.e., defensins 5 and 6 (HD5, HD6) (Jones & Bevins, 1992, 1993). The function, activity, and structure of defensins have been discussed in depth in several literature reviews (Bevins, 2013; Clevers & Bevins, 2013; Lehrer & Lu, 2012; Ouellette, 2011).

2. THE INTESTINAL CRYPT STEM CELL NICHE

2.1 The Discovery of the Intestinal Stem Cell

The crypts of Lieberkühn were for long time suspected to harbor the elusive intestinal stem cells responsible to preserve homeostasis in one of the most dynamic and regenerative epithelial tissues. Indeed, seminal studies by Cheng and Leblond identified cycling cells located at the very bottom of the intestinal crypt, the so-called crypt base columnar cells or CBC's, and proposed them as the stem cells of the gut epithelium. Experimental evidence for the identity of CBCs as stem cells of the adult gut was provided by Barker in the laboratory led by Clevers, based on the identification of the transmembrane receptor *Lgr5* (leucine-rich repeat-containing G protein-coupled receptor 5) as a specific marker of cycling CBCs. Lineage tracing experiments were performed with a knock-in mouse model expressing both a green fluorescent protein (GFP) and a tamoxifen inducible C-recombinase under the control of the endogenous *Lgr5* gene promoter (*Lgr5*-eGFP-IRES-CreER^{t2}) (Barker et al., 2007). These mice, when bred with Rosa26-LacZ reporter animals and induced by tamoxifen, allow the excision of the roadblock (i.e., the stop sequences flanked by LoxP sites located upstream of the LacZ gene) and transcription of the reporter gene by Cre protein expression only in *Lgr5*$^+$ cells. In this way, LacZ expression, easily detectable by X-gal staining, will mark *Lgr5*$^+$ cells and their cellular progeny. Upon tamoxifen treatment, *Lgr5*-eGFP-IRES-CreER^{t2} mice gave rise to long ribbons of blue-labeled cells starting from the bottom of the crypt to the villus tip. These blue ribbons persisted in the intestine of the mice for more than 300 days and encompassed all the intestinal cell lineages, thus showing that *Lgr5*$^+$ have long-term self-renewal and multipotent differentiation capacity both in the small and large intestine.

Although stem cells have long been thought to divide infrequently, the intestinal epithelium almost entirely renews its cellular composition every 5 days suggestive of an actively proliferating stem or precursor cell. By using the above *Lgr5* knock-in mice, CBCs have been shown to divide approx. every 16 h (Snippert et al., 2010). Another common view on stem cells is relative to their mitotic division modality, i.e., in symmetric vs asymmetric fashion giving rise to two daughter cells with equal (two stem or two committed cells) or distinct (one stem cell and one committed progenitor) cell fates, respectively. By employing *Lgr5*-eGFP-IRES-CreER^{t2} mice bred with the Rosa26-confetti reporter, it was shown that *Lgr5*$^{+}$ stem cells divide symmetrically and that they follow a neutral drift dynamic (Snippert et al., 2010).

Several studies suggested that stem cells in epithelial tissues come in two flavors: rapidly and slowly cycling (Fuchs, 2009). *Lgr5*$^{+}$ cells divide every 16 h and as such represent frequently cycling stem cells thought to underlie intestinal homeostasis under physiological conditions. A more quiescent or slowly cycling stem cell type located at position +4 from the crypt base was first identified by Potten in 1974 based on its label-retaining properties (Potten, Kovacs, & Hamilton, 1974). Only several years later the first gene marker was identified to earmark the +4 label-retaining cells, namely, the *Bmi1* (B cell-specific Moloney murine leukemia virus integration site 1) gene. By following a lineage tracing approach analogous to that employed for the *Lgr5*$^{+}$ CBCs, *Bmi1*$^{+}$ cells at the +4 position were shown to give rise to all the differentiated lineages of the small intestine (Sangiorgi & Capecchi, 2008). Additional genes have also been proposed to specifically mark a subset of slowly dividing stem cells such as *mTert*, *Hopx*, and *Lrig1* (Carlone & Breault, 2011; Powell et al., 2012; Takeda et al., 2011). Although lineage tracing provides valuable information on specific cell types and their progeny, it mainly relies on the specificity of the gene promoter under which the Cre recombinase is driven. In situ hybridization analysis on single RNA molecules indicated that none of the aforementioned (+4) markers is uniquely expressed at a single position along the crypt–villus axis (Munoz et al., 2012). However, this is likely to be true for the vast majority of the genes/markers expressed by stem and progenitor cells in the lower intestinal crypt. Most recently, the laboratory led by Eduard Batlle identified a specific RNA-binding protein called Mex3a that, in combination with Lgr5, marks a specific population of quiescent stem cells resident in the lower crypt also located at around position +3/+4 (Barriga et al., 2017). Single cell RNAseq analysis of individual *Lgr5*$^{+}$ stem cells revealed the presence two populations differing by their cycling rates with *Mex3a* expression earmarking the slowly cycling *Lgr5*$^+$ subpopulation. Lineage tracing analysis using *Mex3a* knock-in mice bred with the Rosa26 reporter line conformed that *Mex3a*$^+$/*Lgr5*$^+$ cells represent bona fide infrequently dividing stem cells in vivo (Barriga et al., 2017).

The presence of quiescent stem cells is of functional relevance as they have been proposed to underlie the regenerative response by compensating for the loss of rapid-cycling *Lgr5*$^+$ CBCs upon tissue insults (Yan et al., 2012). In the small intestine, tissue regeneration upon injury conditions has been shown to rely on the dedifferentiation of terminally committed cells. Terminally differentiated cells have long been thought to be unable to rewire their cell fate and reenter the cell cycle and acquire pluripotency. Recently, however, distinct differentiated lineages of the small intestine, i.e., PCs and *Alpi*$^+$ enterocytes, have been shown to act as stem/progenitor-like cells upon tissue injury (Roth et al., 2012; Tetteh et al., 2016). The issue of plasticity as an intrinsic property of specific intestinal cell lineages raises several questions on the underlying pathways and molecular cues which are of relevance for our understanding of analogous mechanisms in pathological conditions such as tumor plasticity and resistance to therapy.

2.2 PCs and Crypt Development

Both the small and large intestinal epithelium develop by tubulogenesis from the posterior endoderm following extensive folding (Noah, Donahue, & Shroyer, 2011; Zorn & Wells, 2009). The embryonic gut tube is lined by a simple epithelium that condenses to form a pseudostratified epithelium where all the cells are attached to the basement membrane. Around e14, the gut epithelium assumes a columnar form, while it forms protruding structures called villi. Lineage specification is initiated at around e17 in proliferating cells confined in the intervillus region from where they subsequently invaginate to form intestinal crypts. Notably, whereas in the mouse crypt development occurs at early postnatal stages and is completed by the time of weaning, in man it occurs in utero (Darmoul, Brown, Selsted, & Ouellette, 1997).

Notwithstanding the absence of fully mature and granulated PCs in the mouse intestinal epithelium during the first 2 postnatal weeks, expression of defensins 1 and 6 was found in scattered fashion throughout the newborn intestinal lining thus raising the possibility of an innate immune role for these peptides at early perinatal stages (Darmoul et al., 1997).

De novo crypt formation occurs through a process called "crypt fission," a rate-limiting event in intestinal expansion and growth that consists in its essence of the division of a crypt into two daughter crypts (Clarke, 1972; St Clair & Osborne, 1985). Crypt fission represents a form of epithelial tube branching through which the newly formed crypts of Lieberkühn divide and multiply thus extending and widening the intestinal tract during postnatal development. The lumen of the resulting two daughter crypts can be of equal or unequal length as the result of symmetric and asymmetric crypt fission, respectively. However, the cellular and molecular mechanisms underlying crypt fission are still poorly understood.

Recently, Inke Nathke and collaborators observed that the number and relative position of PCs and *Lgr5*$^{+}$ CBCs at the bottom of the crypt are critical for the initiation of crypt fission (Langlands et al., 2016). Analysis of crypts undergoing fission revealed that the process can be divided into an early and a late phase (Langlands et al., 2016). During early fission, a change in cell patterning occurs with PCs absent from the middle of the crypt bottom due to their migration to either side of the site of bifurcation/branching. Simultaneously to this side clustering of PCs, *Lgr5*$^{+}$ CBCs form a distinct cluster in the middle of the crypt base thus marking the bifurcation site where, during the late fission phase, the epithelial lining is expanded upward until the duplication of the two daughter crypts is completed (Figure 1A).

The specific cell patterning observed during crypt fission reflects the intrinsic cell properties of *Lgr5*$^{+}$ CBCs and PCs, the latter showing high stiffness and increased adhesion to the basement membrane. High β4 integrin expression allows PCs to firmly anchor to the basement membrane when compared with other cell types.

In confirmation of previous studies showing that inhibition of ephrin (Eph) signaling causes misplacement of PCs from the crypt base (Batlle et al., 2002), inhibitory Eph protein fragments have been found to induce their mislocalization in ex vivo organoids though without altering the rate of crypt fission. However, altering PC localization skewed the balance between symmetric and asymmetric fission in favor of the latter, likely due to the formation of new branches further away from the crypts. Cell adhesion and integrin levels appear to play a central role in this process as shown by the observed 50% reduction in crypt fission in the presence of a β4 integrin antibody. It is therefore plausible to postulate that the enhanced adhesion, a feature of PCs when compared to their neighboring cells, plays a key functional role in the early phase of crypt fission (Langlands et al., 2016).

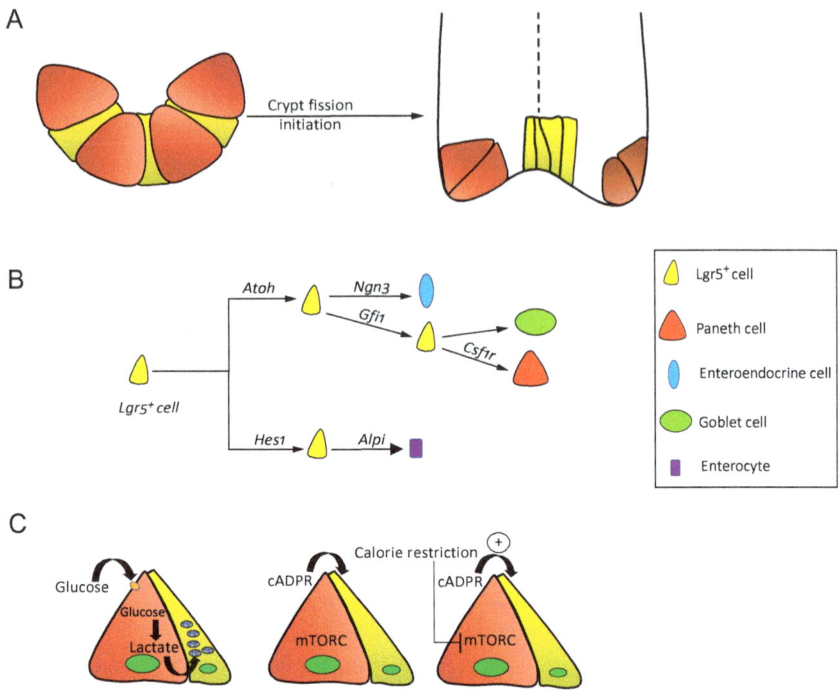

Fig. 1 (A) At the bottom of the crypt of Lieberkühn, Paneth cells (*brown*) and *Lgr5*$^+$ stem cells (*yellow*) cluster away upon crypt fission initiation (Langlands et al., 2016). The increased stiffness of Paneth cells forms harder regions on the sides of the branching crypt, in contrast with the "softer" Lgr5$^+$ stem cells positioned in the center. (B) Schematic diagram of the differentiation fates of the *Lgr5*$^+$ stem cell and the genes that underlie the specific commitment steps. (C) *Left*: Paneth cells (*brown*) sustain the metabolism of *Lgr5*$^+$ cells by using glucose as a substrate to secrete lactate that, during homeostasis, fuels oxidative phosphorylation in *Lgr5*$^+$ stem cells. *Center* and *right*: mTORC1 inhibition in Paneth cells leads to the secretion of cyclic ADP ribose (cADPR). During caloric restriction, mTORC1 is inhibited in Paneth cells leading the enhanced cADPR secretion and increased *Lgr5*$^+$ stem cell self-renewal and multiplicity.

2.3 Paneth-Specific Signaling Pathways: Wnt and Notch

Experimental evidence links de novo crypt formation to canonical Wnt signaling in view of the involvement of two of its downstream target genes (*EphB* and *cMyc*) in the migration and compartmentalization of Paneth and *Lgr5*$^+$ cells within the lower crypt. Cell migration within the crypt occurs in bidirectional fashion: whereas enteroendocrine cells, goblet cells, and enterocytes migrate upward along the crypt–villus axis, PCs descend in opposite direction toward the bottom of the crypts. In the mouse prenatal intestine, the EphB2 and EphB3 receptors and their ligand Ephrin-B1 are expressed in complementary domains: both receptors and ligand are coexpressed in cells located at the periphery of the intervillus pockets, whereas expression of the EphB2 and EphB3 earmarks the cells present at the very bottom of the pockets (Batlle et al., 2002). In the adult small intestine, the expression pattern of the two ephrin receptors appears to be more complex with EphB3 being restricted to PCs in the crypt bottom, whereas *EphB2* expression earmarks the entire crypt bottom until position +4.

EphB3 receptor knockout mice are characterized by misplaced PCs throughout the crypt–villus axis, a phenotype not observed in *EphB2*$^{-/-}$ mice. Therefore, expression of the EphB3 receptor in PCs appears necessary for their correct positioning at the bottom crypt (Batlle et al., 2002). Furthermore, the nuclear β-catenin accumulation characteristic of PCs at the bottom of the crypt was lost in their misplaced equivalents in *EphB3*$^{-/-}$ mice thus indicating a noncell autonomous and localization-dependent regulation of Wnt signaling in PCs (Batlle et al., 2002). As previously mentioned, Wnt signaling plays a pivotal role in the intestinal epithelium with both *Lgr5*$^{+}$ CBCs and PCs are characterized by the accumulation of β-catenin in the nucleus where it binds with members of the family of Tcf transcription factors thus driving the expression of Wnt target genes (van de Wetering et al., 2002).

Cytosolic levels of β-catenin are tightly controlled by a multiprotein complex encompassing the tumor suppressors Apc (adenomatous polyposis coli) and Axin, the Ser/Thr kinases GSK3β (glycogen synthase kinase 3-beta) and CK1 (casein kinase 1), protein phosphatase 2A (PP2A), and the E3-ubiquitin ligase β-TrCP. In the absence of Wnt ligands, this "destruction complex" phosphorylates β-catenin promoting its ubiquitinilation and degradation by the proteasome (Stamos & Weis, 2013). Of note, canonical Wnt/β-catenin signaling has been shown to induce maturation of PCs in the mouse small intestine: expression of the Paneth-specific gene program is lost in the intestine of *Tcf4*$^{-/-}$ embryos (van Es et al., 2005). In particular, failure to express defensins and cryptdins caused PCs maturation defects. Also, the frizzled (Fzd) family of receptors encompasses integral membrane proteins featuring seven - transmembrane-spanning domains function in canonical Wnt signaling. Remarkably, *Fzd5* knockout mice exhibit an identical phenotype to that of *EphB3*$^{-/-}$ animals with misplaced PCs with no β-catenin nuclear accumulation, in confirmation of the positional cues provided by Wnt signals along the crypt–villus axis (Batlle et al., 2002; van Es et al., 2005). The Wnt-driven maturation of PCs and their pronounced nuclear β-catenin accumulation is intriguing in view of the essential role played by canonical Wnt signaling in the regulation of stemness in the intestinal crypt. Hence, the same pathway controls the onset and maintenance of both multipotent and fully differentiated (and postmitotic) lineages.

Studies aiming at the identification of the Wnt signals required for PC maturation and maintenance have been performed by first establishing the expression patterns of Wnt ligands in the intestinal epithelium (i.e., Wnt3; uniquely expressed in PCs) and in the surrounding mesenchyme (Wnt2b, Wnt4, and Wnt5a). Notably, conditional ablation of Wnt3 throughout the intestinal epithelium (*Wnt3*$^{fl/fl}$/*Vil*^{CreERT2}) does not affect PCs, indicative of functional redundancy between mesenchymal and epithelial Wnt ligands in vivo (Farin, Van Es, & Clevers, 2012). The latter was confirmed (i) by the impaired growth of ex vivo cultured crypt organoids derived from the *Wnt3*$^{fl/fl}$/*Vil*^{CreERT2} mice upon removal of the floxed *Wnt3* allele and (ii) by the ability of exogenously added Wnt3 to rescue the growth impairment (Farin et al., 2012).

The orphan G protein-coupled receptor Lgr4 is expressed in small intestinal crypts above the PC zone (i.e., in the TA region), and in CBCs and in a subset of PCs (Mustata et al., 2011). *Lgr4*$^{-/-}$ mice show reduced epithelial proliferation without any significant defect in the differentiation of absorptive, enteroendocrine, and goblet cell lineages. However, the absence of specific PC markers such as lysozyme and cryptdin 4 was also noticed in *Lgr4*$^{-/-}$ mice, likely indicative of a PC maturation defect. Accordingly, ex vivo cultured crypts from the *Lgr4* knockout mice failed to form fully mature organoids and showed decreased expression levels of several stem cell-specific markers and Wnt targets such as *Axin2*, *Sox9*, and *Lgr5*. Accordingly, reactivation of Wnt signaling by addition of lithium chloride (LiCl) to the organoid culture medium partially rescued their growth impairment (Mustata et al., 2011).

Among the Wnt targets observed to play a key role in PC onset, *Sox9* is expressed before the appearance of PCs in the embryonal gut and its expression overlaps with that of cryptdins in the intervillus pockets. Inducible genetic ablation of *Sox9* results in the disappearance of PCs and a decrease of goblet cells (Bastide et al., 2007; Mori-Akiyama et al., 2007). As a consequence of the altered tissue morphology and absence of PCs, the crypts of *Sox9*$^{fl/fl}$ mice present an enlarged proliferative compartment extended to the whole crypt bottom (Bastide et al., 2007; Mori-Akiyama et al., 2007).

Constitutive *Sox9* expression as a consequence of loss of *Apc* function has also been shown to induce formation of ectopic PCs in the mouse colon (Feng et al., 2013). By employing a colon-specific promotor (*Cdx2*) to delete the *Apc* tumor suppressor gene, it was shown that adenoma formation is accompanied by the appearance of ectopic lysozyme-expressing Paneth-like cells. These metaplastic PCs were observed in regions outside of the crypt base, often at new crypt budding sites, coincident with high *Sox9* expression as the result of the induced *Apc* deletion. The latter is not surprising in view the role of *Sox9* as Wnt target, but it is noteworthy that *Sox9* upregulation in the colon of these mice preceded the changes in β-catenin subcellular localization possibly due its easier detection, when compared to β-catenin stabilization, at early time points after *Apc* inactivation in the colon epithelium (Feng et al., 2013).

Alongside canonical Wnt, also Notch signaling plays an important role in PC differentiation and maturation, mainly through one of its downstream target genes, namely *Math1*, encoding for a basic helix–loop–helix transcription factor. Studies performed on *Math1* knockout mice revealed the ablation of all secretory lineages including Paneth and goblet cells, and enteroendocrine cells (Yang, Bermingham, Finegold, & Zoghbi, 2001). Thus, *Math1* is required for progenitor cells to commit toward a secretory fate. Notably, studies performed on mosaic *Math1*-null crypts showed that commitment toward a secretory fate improves regeneration upon small bowel resection (SBR). *Math1*-deficient crypts showed decreased regenerative capacity after SBR when compared to proficient crypts, suggesting that secretory lineages may play a (niche) role in the response to tissue injury (Shroyer et al., 2007).

Constitutive expression of an active form of the Notch 1 receptor in the intestinal epithelium causes signs of malnourishment at postnatal day 1 (P1) and death at P3 as a consequence of the inhibition of all secretory lineages, a phenotype opposite to that observed in *Hes1* knockout mice characterized by an excess of secretory cells at the expenses of enterocytes (Fre et al., 2005). Hence, constitutive Notch 1 expression upregulates *Hes1* expression while downregulating *Math1*, a key gene in secretory lineage development.

Notch signaling inhibition has also been studied: mice treated with dibenzazepine showed an increment in PCs numbers, with abnormal lysozyme staining, i.e., cytoplasmic and diffused rather than in the characteristic granular pattern, indicative of alterations in PCs formation upon Notch signaling inhibition (VanDussen et al., 2012). Upon further analysis, both the number and size of PC granules were shown to be decreased in addition to the expression of goblet-specific markers, suggesting the establishment of an intermediate phenotype in between the Paneth and goblet lineages upon Notch signaling inhibition. These findings highlight the importance of Notch signaling in terminal differentiation of PCs (VanDussen et al., 2012). Screens performed on *Math1*-proficient and -deficient mice led to the identification of the target genes of this transcription factor critical for the establishment of the secretory cell fate (Shroyer, Wallis, Venken, Bellen, & Zoghbi, 2005). The *Gfi1* gene was found to be significantly downregulated upon *Math1* ablation. Likewise, *Gfi1* knockout mice were deprived of mature PCs together with a significant reduction of goblet cells and an increase in enteroendocrine cells. In these mice, secretory progenitors although already committed, are unable to differentiate into mature PCs (Figure 1B) (Shroyer et al., 2005).

Another Paneth-specific gene known to affect Notch signaling in the crypt bottom encodes for the receptor of colony stimulation factor 1 (*Csf1r*), whose *Csf1* ligand is expressed by cells in close proximity to PCs. Csf1 receptor-deficient mice (*Csf1r*$^{-/-}$) are characterized by a marked reduction in fully mature PCs (Huynh et al., 2009, 2013). Goblet cell multiplicity was also highly increased as a consequence of the *Csf1r* genetic ablation indicative of faulty cell fate determination in the common goblet and PC precursor. These results were also confirmed in ex vivo cultured crypts, where *Csf1r*$^{-/-}$ organoids showed decreased clonogenicity and size, and reduced number of budding events. These experiments also suggested a role in PC maintenance for the Csf1 receptor: conditional *Csf1r* deletion by villin-Cre in organoids resulted in PCs loss (Figure 1B) (Akcora et al., 2013).

In view of the main secretory function of PCs, it is not surprising that proteins involved in vesicle trafficking and secretion play significant roles in PC maturation. In particular, two classes of vesicle proteins have been investigated in gut epithelial homeostasis, namely, Rab8A and αSNAP (soluble *N*-ethylmaleimide-sensitive factor attachment protein alpha). The Rab family of small GTPases regulates both the sorting of transported molecules to the correct vesicles and vesicle delivery to target membranes. Of note, Rab8A has been linked to trafficking and secretion of Wnt ligands as it mediates anterograde transport of Gpr177 (wntless), a Wnt-specific transmembrane transporter.

Rab8a$^{-/-}$ mice lack fully mature PCs and are instead characterized by immature cells with decreased numbers of granules (Das et al., 2015). Expression analysis of the intestinal epithelium of *Rab8a*$^{-/-}$ mice showed downregulation of several Wnt targets such as *Axin2* and *Ascl2* together with decreased nuclear β-catenin levels and reduced lysozyme and cryptdin 5 expression. When cultured ex vivo, crypts from *Rab8a*$^{-/-}$ mice showed less budding events and decreased clonogenicity. Accordingly, the negative effect brought about by *Rab8A* ablation can be rescued by adding Wnt3 in the organoid medium. The observed effect on cell maturation seems to be specific for PCs and Wnt secretion, as goblet cell numbers, mucin trafficking, packaging, and secretion were unaltered (Das et al., 2015).

Trafficking and delivery to target membranes is followed by vesicle fusion that requires the SNARE complex, the postfusion disassembling of which is regulated by the aforementioned factor αSNAP. A pronounced reduction in PC differentiation was reported in ex vivo cultured crypts derived from mice carrying a single amino acid substitution in the *αSNAP* gene leading to its decreased expression in the intestine (Lechuga, Naydenov, Feygin, Jimenez, & Ivanov, 2017). These results point to a selective inhibition of terminal differentiation of PCs due to the *αSNAP* gene defect and to a novel functional role for vesicle trafficking and fusion in the regulation of secretory cell fate.

2.4 PCs Constitute the Main Epithelial Intestinal Stem Cell Niche

The physical association of PCs and *Lgr5*$^{+}$ CBCs at the bottom of the crypt is suggestive of an additional function for this multitasking secretory lineage, namely, as niche cells capable of regulating the activity of stem cells in homeostasis and during regeneration upon tissue insults. Sato and collaborators in the Clevers' laboratory noticed that *Lgr5*$^{+}$ stem cells were extremely inefficient when employed to establish ex vivo organoids unless cocultured with PCs, which increased organoid multiplicity by at least 20-fold (Sato et al., 2011). Based on this seminal study, PCs were proposed to represent the main epithelial niche for *Lgr5*$^{+}$ stem cells.

As discussed earlier, Wnt and Notch signaling play rate-limiting roles in regulating both stemness and in establishing secretory lineage fate; likewise, a tightly controlled balance between the two pathways is crucial to regulate intestinal homeostasis (Tian et al., 2015). Specific Wnt and Notch factors encoded by PCs, namely, Wnt3, Dll1, and Dll4 (delta-like ligands 1 and 4), underlie their niche function (Sato et al., 2011). The organoid assay (further discussed later) and the possibility of culturing these two cell types ex vivo have been instrumental for the identification of these and other key regulatory niche cues (Sato et al., 2009).

Due to their close physical association with *Lgr5*$^+$ cells, PCs exert their niche function both through secreted short-range factors and by membrane-bound signaling ligands. Moreover, apart from their signaling modalities, PCs provide additional support to *Lgr5*$^+$ stem cells in a contact-dependent manner (Gracz et al., 2015). Although these requirements are well established for rapidly cycling *Lgr5*$^+$ CBCs, little is known about the niche requirements of other stem cell types, e.g., the more quiescent stem cells located at position +4.

Additional stem cell niche factors are thought to be secreted by the surrounding stroma. In particular myofibroblasts have been shown to secrete Wnt ligands (Vermeulen et al., 2010). Studies on Porcupine (*Porcn*), a gene encoding for an *O*-acetyltransferase essential for Wnt ligands secretion, showed that its genetic ablation in epithelial cells does not impair intestinal homeostasis and regeneration upon tissue insults. However, concomitant genetic *Porcn* ablation and administration of a specific porcupine inhibitor (C90) leading to impaired Wnt secretion in both the epithelial and stromal compartment significantly reduced *Lgr5*$^+$ CBCs multiplicity in homeostasis and stalled proliferation and recovery after tissue injury (Kabiri et al., 2014). These studies highlight the relative stromal vs epithelial contribution in the secretion of Wnt and other ligands known to regulate intestinal stemness and must be taken into consideration when addressing the in vivo niche role of PCs. Accordingly, in vivo depletion of PCs, as observed in *Sox9* knockout mice, does not results in any major differentiation defect among the different intestinal epithelial cell lineages (Mori-Akiyama et al., 2007). Likewise, deletion of the *Atoh1/Math1* gene encoding for a transcription factor necessary for the commitment of secretory lineages results in a loss of PCs without affecting epithelial morphology and homeostasis (Durand et al., 2012). Similar results were obtained by knocking out other genes (e.g., *Gfi1*, Shroyer et al., 2005; *Spdef*, Gregorieff et al., 2009). However, it is unclear whether the apparent absence of PCs in these mouse models is limited to fully mature PCs or is extended to its immature progenitors which could still exert some of the relevant niche functions. Also, as reported later, PCs are likely to play key roles (both as niche and stem-like cells) in the regeneration of the intestinal epithelium upon inflammation and other forms of tissue injury. As such, mouse models depleted of PCs should be challenged by inflammatory insults to fully evaluate the consequences of their partial or complete loss.

PCs have mainly been studied in inbred C57BL/6J (B6) mice. This is of relevance as expression of PC-specific genes, and pathways may differ among distinct inbred strains and as such differentially affect their secretory, niche, and other functions as illustrated by the case of the secretory phospholipase A2 (*Pla2g2a*). The *Pla2g2a* gene, specifically expressed in PCs, represents a major genetic modifier of intestinal tumor multiplicity in the Apc^{Min} mouse model (Dietrich et al., 1993; MacPhee et al., 1995). While the C57BL/6J strain carries a *null* secretory phospholipase A2 allele ($Pla2g2a^{-/-}$) due to a stop codon mutation, other inbred strains (e.g., FVB, AKR, BALB/C) are *Pla2g2a* proficient ($Pla2g2a^{+/+}$). When bred into the $Apc^{Min/+}$ background, B6 animals show a strikingly high multiplicity of upper GI adenomas (approx. 90). In contrast, $Apc^{Min/+}/Pla2g2a^{+/+}$ mice are characterized by a pronounced reduction in intestinal tumor numbers. In a recent study, our laboratory has shown that the intracellular pool of phospholipase A2 downregulates Wnt signaling during homeostasis by modifying the subcellular localization of the yes-associated protein 1 (Yap1) protein (Schewe et al., 2016). Upon inflammation, Pla2g2a is secreted into the lumen from where it activates a cascade of downstream signaling events culminating in the synthesis of prostaglandins and Wnt activation, thus supporting the regenerative response to the inflammatory insults (Schewe et al., 2016).

Overall, it appears that PCs exert their niche function throughout a complex and diverse network of secretory auto- and paracrine pathways and by their physical association with stem cells.

2.5 PCs and Nutrient Sensing: A Matter of Metabolism?

As mentioned earlier in this chapter, already the first studies on PCs observed a change in their content upon feeding mice with different diets (Cowdry, 1934; Georges, 1930) thus suggesting a functional connection with metabolism.

Although a set of defined factors including Notch and Wnt stimuli have been proposed to constitute the stem cell niche (Sato et al., 2011), very little is known about the metabolic requirements of $Lgr5^+$ CBCs and their niche. More recently, analysis of Paneth and $Lgr5^+$ cells revealed a striking metabolic dichotomy: while PCs are earmarked by high glycolytic activity, $Lgr5^+$ CBCs rely on mitochondrial oxidative phosphorylation for their metabolic needs (Rodriguez-Colman et al., 2017). Notably, PCs, characterized by a high glycolytic activity, secrete lactate which is taken up by the $Lgr5^+$ stem cells to fuel oxidative phosphorylation. As such, PCs provide a metabolic niche to $Lgr5^+$ stem cells.

Other studies involving dietary modulation suggested that PCs play a major role in sensing the organism's nutritional status. In particular, Yilmaz et al. showed that, upon longevity-promoting calorie restriction (CR), PCs elicit *Lgr5*$^+$ stem cell function by downregulating mTORC1 (mechanistic target of rapamycin complex 1) signaling (Yilmaz et al., 2012). Of note, CR seems to primarily affect PCs but not CBCs.

mTORC1 inhibition in PCs leads to the activation of the ectoenzyme bone stromal antigen 1 (Bst1) which in turn triggers the secretion of cyclic ADP ribose (cADPR). During caloric restriction, this paracrine factor is secreted by PCs and favors stem cell self-renewal eventually resulting in an increase of *Lgr5*$^+$ CBCs' multiplicity. This effect could be mimicked in vivo by administration of rapamycin, the main mTORC1 pharmacological inhibitor. However, in a subsequent study, Igarashi and Guarente reported a more detailed analysis of the effects of CR with mTORC1 signaling being inhibited in PCs but upregulated in intestinal stem cells (Igarashi & Guarente, 2016). In this scenario, mTORC1 activation is mediated by SIRT1 in *Lgr5*$^+$ cells leading to an increase in protein synthesis and an increase in CBC multiplicity. Notably, in this study the in vivo administration of rapamycin was shown to abolish CBC expansion rather than mimicking CR effects (Igarashi & Guarente, 2016). Although a model was proposed where PC signaling could override any direct nutrient sensing in *Lgr5*$^+$ CBCs, further studies will be required to elucidate how drugs that modulate CR pathways may exert opposing effects on different cell types (Figure 1C).

2.6 Culturing Intestinal Crypts ex vivo as Organoids

Seminal work by the laboratories led by Clevers and Kuo's led to the establishment of novel 3D culture methods for the ex vivo culture of mouse intestinal crypts as organoids or "miniguts," i.e., long-lived and self-renewing structures recapitulating the crypt–villus organization of the intestine and encompassing both *Lgr5*$^+$ and PCs as the main units of the stem cell niche, in addition to other differentiated lineages (Ootani et al., 2009; Sato et al., 2009). Both methods successfully established long-term epithelial cultures from the small and large intestine by employing matrigel and collagen, respectively. Both studies identified R-spondin, the main ligand of the Lgr5 receptor and a potent canonical Wnt signaling amplifier, as the key growth factor for the successful organoid culture. Other important factors include the epithelial growth factor and the Tgf-β inhibitor Noggin.

Organoid cultures are also partly dependent on the Wnt3A ligand, normally present in the basolateral membrane of PCs (Farin et al., 2016). Recently the combination of Wnt3A, R-spondin, and Noggin (WRN) has been found to be sufficient to grow ex vivo intestinal organoids from most companion and large animals (Powell & Behnke, 2017).

The availability of an ex vivo culture system that, although in the absence of a mesenchymal/stromal niche support, recapitulates the complex in vivo structure of the intestine, stimulated and improved additional studies through the genetic and biochemical modifications of the whole crypt or its specific cell lineage components (Andersson-Rolf et al., 2016, 2017; Koo et al., 2011; Rodriguez-Colman et al., 2017; Schewe et al., 2016).

As organoids can also be established from patient-derived tumors, biobanks of human 3D cultures are currently being established from tumor material and matched healthy tissues for several purposes including the pre-clinical assessment of (chemo) therapeutic approaches, and studies of stem cell tracing and plasticity within neoplastic lesions, tumor heterogeneity, and metabolism. The establishment and in vitro maintenance of organoids from patient-derived colon cancers require different and specific factors depending on the tumor subtype, also in reflection of the heterogeneity of the spectrum of malignancies affecting the large bowel (Fujii et al., 2016).

As mentioned earlier, the main limitation of the currently available organoid culture methods is the lack of an intact stromal microenvironment. Also, since in its original formulation the protocol employs whole intestinal crypts as the biological substrate to establish organoids, this precludes functional studies of specific cell lineages within the stem cell niche, and in particular its two main components, namely, the *Lgr5*$^+$ CBCs and PCs.

In order to include and preserve the stromal and extracellular matrix microenvironment where intestinal crypts normally reside, colonic stroma has been de- and recellularized with myofibroblasts, endothelial cells, and epithelial cells (whole organoids or sorted cells) (Chen et al., 2016). This approach promises to provide additional insights into the normal gut physiology, not currently feasible with the canonical organoid culture protocol.

As for the second limitation, *Lgr5*$^+$ stem cells and PCs tend to physically interact when coincubated to then give rise to organoids (Sato et al., 2011). This key feature is the basis for the "organoid reconstitution assay" (ORA) further developed and implemented by us and others (Igarashi & Guarente, 2016; Schewe et al., 2016). Sorting by FACS of CBCs and PCs allows their genetic and/or biochemical modification (or their isolation from genetically modified or treated mice) prior to their coincubation and reconstitution into organoids, thus providing a handy functional study tool. A recent example of the usefulness of ORA is represented by the demonstration of the striking metabolic dichotomy between PCs (glycolytic) and *Lgr5*$^+$ stem cells (oxidative phosphorylation) and the functional relevance of these metabolic needs for the maintenance of homeostasis (Rodriguez-Colman et al., 2017).

3. PCs, DYSBIOSIS, AND IBD

The intestinal microbiome hosts a diverse and abundant group of microorganisms known to be essential for gut homeostasis and development, as shown by studies on germ-free mice. Several environmental factors affect the composition of the microbiome such as diet (Claesson et al., 2012; Wu et al., 2011), oxygen levels, and pH (Bevins & Salzman, 2011). In the laboratory mouse, inbred strain-specific variations in the composition of the microbiome have been shown (Gulati et al., 2012), analogous to the observed variations in PCs' multiplicity, distribution, and secretory function (Salzman & Bevins, 2013). Dysbiosis, i.e., the unfavorably altered composition of the microbiota, is central to a broad spectrum of human pathologies, among other IBD.

The term IBD is comprehensive of two main chronic and relapsing pathologies of the intestinal tract, namely, UC and CD (Kaser, Zeissig, & Blumberg, 2010). Like many other pathological conditions such as diabetes, atherosclerosis, obesity, and cancer, IBD is nowadays thought to result from inappropriate inflammatory responses to dysbiosis in genetically susceptible individuals (Figure 2A) (Salzman & Bevins, 2013). One of the main differences between UD and CD is the affected region of the gastrointestinal tract. UC is characterized by chronic inflammation of the colon, initiating in the rectal region, and subsequently spreading proximally, with tissue damage usually limited to the mucosa or submucosa featuring local crypt inflammation (*cryptitis*) and abscesses with neutrophilic exudate in the lumen. CD instead affects both the large bowel and the distal part of the small intestinal tract, namely, the ileum where PCs are more abundant. At the microscopic level, CD's distinctive features include thickening of the submucosa, transmural inflammation, fissuring ulceration, and granulomas (Kaser et al., 2010).

As a consequence of the primary role played by PCs in the secretion of AMPs and in the control of the gut microbiota during homeostasis, their dysfunction (either due to genetic defects or to environmental "stressors" such as infections, obesity, and graft vs host disease) has been shown to be

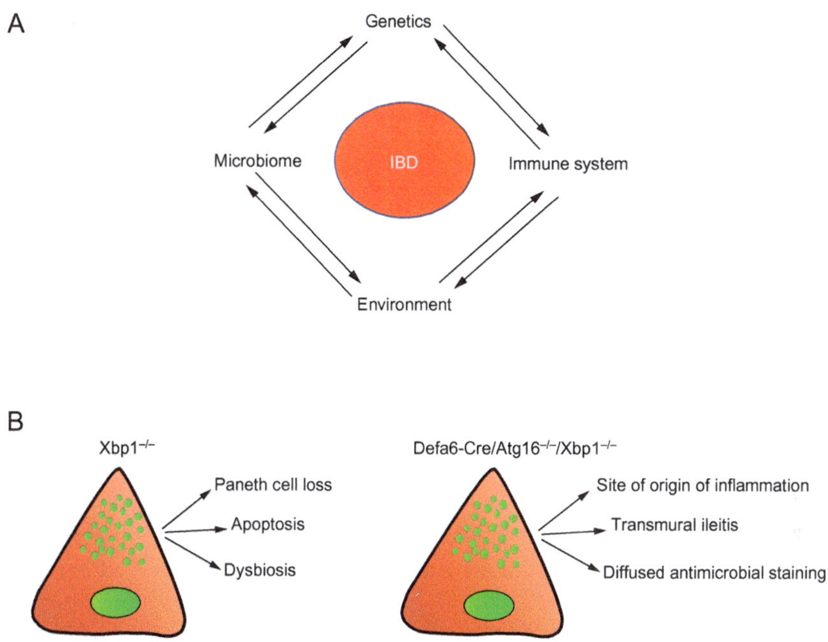

Fig. 2 (A) Inflammatory bowel disease (IBD) is a complex multifactorial disease dominated by four major cross-interacting components, namely the genetic background of the host, the gut microbiome, the host's immune system, and the environment. (B) Genetic depletion of the *Xbp1* gene encoding for a key ER stress response protein causes Paneth cell loss, increased apoptosis, and dysbiosis. Combined genetic loss of the *Xbp1* and *Atg16* genes in Paneth cells results in diffused lysozyme staining reminiscent of IBD and transmural ileitis, suggesting that PCs represent the site of origin of intestinal inflammation (Adolph et al., 2013).

causative in a broad spectrum of dysbiosis-associated pathologies, including IBD (Salzman & Bevins, 2013). The complex multifactorial nature of IBD can be in fact reduced to the interaction of a quartet of host-derived and environmental factors: the genetic predisposition and susceptibility of the host, the intestinal microbiota, the immune system, and the above-mentioned environmental stressors (Clevers & Bevins, 2013).

Genetic susceptibility predisposes the individual to IBD and, likewise, family history does represent a risk factor as shown by genome-wide association studies (GWAS). A multitude (>150) of IBD-associated single nucleotide polymorphisms (SNPs) have been established as risk loci, 28 of which common to both CD and UC. Of note, the very first genes to be identified are not only highly and specifically expressed in PCs but also exert essential roles in PC's secretory function (Cho, 2008; VanDussen et al., 2014) including host-bacterial interaction and response (e.g., *NOD2*) (Hugot et al., 2001; Ogura et al., 2001), ER stress and unfolded protein response (*XBP1*) (Kaser et al., 2008), and autophagy (*ATGL16L1*) (Cadwell et al., 2008; Cadwell, Patel, Komatsu, Virgin, & Stappenbeck, 2009).

3.1 Nod2 in IBD and PCs

Although its physiological function in the intestine remains elusive, the *Nod2* gene, encoding for a member of the nucleotide-binding oligomerization domain–leucine-rich repeat (NOD-LRR) family of proteins, is specifically expressed in PCs and is most pronounced in the ileum of both healthy controls and IBD patients (Ogura et al., 2003). A role in bacterial sensing through innate and adaptive immunity was initially hypothesized and later confirmed by the laboratory of Richard Flavell (Kobayashi et al., 2005). Strikingly, PCs from *Nod2*-deficient mice were not able to sense specific microbial antigens such as muramyl dipeptide (MDP), a conserved bacterial PGN. Nevertheless, $Nod2^{-/-}$ mice did not display any overt symptoms of intestinal inflammation and were not significantly susceptible to DSS-induced colitis. However, they were resistant to lipopolysaccharide (i.e., Toll-like receptor agonists) challenge with MDP priming, in confirmation of the role of *Nod2* in sensing and responding to bacterial antigens by activating the adaptive immune system either directly or by enhancing the production of α-defensins (Kobayashi et al., 2005). As such, Nod2 is essential to protect the host from intestinal bacterial infection. However, it still remains to be established whether the PC defect in MDP sensing is the only and even main mechanism through which *Nod2* mutations are related to CD development. Rather than representing the main initiating factor for the disease, *Nod2* gene defects might alter the physiological response to pathogenic bacteria and predispose the individual to CD.

Additional studies in germ-free mice revealed that lack of lysozyme expression also characterizes PCs from $Nod2^{-/-}$ mice due to specific lysosomal degradation (Zhang et al., 2015). When AMPs are synthetized in PCs, they are sorted into specialized secretory granules called dense core vesicles (DCV). Accordingly, Nod2 is also recruited to AMP-containing DCVs that, in turn, is required for the DCV localization of the multiprotein kinase Lrkk2 and the GTPase Rab2a (Zhang et al., 2015). Defects in the Nod2–Lrkk2–Rab2a axis result in the lysosomal degradation of lysozyme (Zhang et al., 2015).

3.2 ER Stress and Unfolded Protein Response in IBD and PCs

As part of their secretory function, PCs are characterized by a pronounced endoplasmic reticulum and Golgi apparatus, and by large secretory granules. As such, they act as protein factories responsible for the synthesis of many proteins with antimicrobial and essential for intestinal homeostasis. In secretory cells, nascent proteins enter the ER as unfolded polypeptide chains to complete folding and maturation. The folding capacity of the ER of secretory cells is targeted on the rate of nascent proteins entering the ER. Specific sensors monitor the ER lumen and signal to other cellular components. Whenever the rate of unfolded nascent proteins exceeds the folding capacity of the ER, a condition known as ER stress, the unfolded protein response (UPR) is triggered (Ron & Walter, 2007). In this process, the molecular sensor and ER chaperone immunoglobulin-binding protein B (BIP, HSPA5, or GRP78) play a central role. During homeostatic conditions BIP is bound to the luminal side of three master regulators of the UPR response, namely, transmembrane proteins inositol requiring 1 (IRE1), PKR-like ER kinase (PERK), and activating transcription factor 6 (ATF6). Upon ER stress activation, BIP dissociates from these sensors thus triggering their activation. The downstream pathways involve, among others, the Paneth-specific transcription factor X-box-binding protein-1 (XBP1) whose mRNA is converted upon ER stress to an active form by the IRE1 nuclease. This leads to the modulation of protein synthesis to adapt the ER folding capacity to the specific requirements of the cell. Whenever ER stress is not solved, apoptosis is triggered (Ron & Walter, 2007; Schroder & Kaufman, 2005).

PCs, as many secretory cells, are characterized by a basal ER stress level. Of note, the small intestine is the only tissue reported to date to express an isoform (IRE1β) of the proximal sensor IRE1 (Walter & Ron, 2011). Remarkably, in *Xbp1* knockout mice ER stress was detected in the intestinal epithelium through GRP78 upregulation in association with an inflammatory phenotype reminiscent of human IBD with crypt abscesses, ulcerations, and leukocyte infiltration (Kaser et al., 2008). Furthermore, PC with normal granules was barely detectable together with a substantially decreased expression of cryptdins 1, 4, and 5. Also, PC apoptosis, most likely triggered by the failure of solving ER stress, was a feature of *Xbp1* genetic ablation. Last, *Xbp1* knockout mice were more susceptible to infection by gram-positive bacteria and to DSS-induced colitis (Figure 2B).

3.3 Autophagy in IBD and PCs

The term autophagy refers to the natural process through which cellular components, like organelles or large protein complexes which cannot be removed by the proteasome, are degraded and recycled in orderly and regulated fashion through the lysosome. Although the original function of autophagy is likely to have arisen as an adaptation to starvation or nutrient deprivation, it has evolved as a quality control process for organelles and proteins and to regulate energy homeostasis (He & Klionsky, 2009). Moreover, it is also employed by the cell to degrade microorganisms (also referred to as xenophagy), as shown by the increased susceptibility to infections caused by intracellular pathogens upon mutation of autophagy genes (Glick, Barth, & Macleod, 2010). Autophagy can also be selective when it triggers the degradation of specific damaged organelles, such as mitochondria or peroxisomes.

The formation of autophagic vesicles earmarks autophagy (Mizushima, Yoshimori, & Ohsumi, 2011) and is dependent on the translocation of the mTOR substrate complex from the cytosol to the ER. Through recruitment of PI3 kinases to the ER, vesicle formation and elongation are started. The final stage of this process is mediated by a number of autophagy (Atg)-related proteins, namely, Atg12, Atg5, and Atg16, which form a complex with Atg8 and phospholipid phosphoethanolamine, that catalyzes further elongation and maturation of the autophagic vesicles. Once mature, they are transported to lysosomes where they undergo degradation.

The autophagy machinery interfaces with cellular stress and response pathways, including those controlling immune responses and, in turn, inflammation. Large autophagic vesicles were observed to be prominent in PCs after irradiation (Gorbunov & Kiang, 2009). Several mice carrying hypomorphic variants of different *Atg* genes (i.e., *Atg16*, *Atg5*, and *Atg7*) were characterized by apparently mature PCs though with secretory granules decreased in size and numbers (Cadwell et al., 2009). Defects in granules exocytosis leading to decreased AMP secretion into the lumen and differences in microbiome composition were also reported in these mice (Cadwell et al., 2009).

Several defects in autophagy, autophagy-related genes, and PC morphology have been observed in IBD patients (Stappenbeck & McGovern, 2017). As previously mentioned, SNPs in the *ATG* and related genes were found by GWAS to predispose to IBD (Cadwell et al., 2008, 2009; Liu et al., 2017). More recently, a seminal study by Adolph et al. has functionally linked ER stress and autophagy in a specific cell type, the PC (Adolph et al., 2013).

Here, mouse models were developed that were defective in the *Xbp1* gene in the small intestine. Upon induction of ER stress in the intestinal epithelium, autophagy was observed mostly at the bottom of the crypts where PCs reside. To assess whether autophagy was induced to ameliorate ER stress, mice defective in both Xbp1 and Atg16L1 (or Atg7) in intestinal epithelium-specific fashion were generated (Adolph et al., 2013). These mice lacked UPR-induced autophagy resulting in constitutive ER stress in the absence of autophagy. Notably, the intestines of the double-mutant mice featured a strong inflammatory reaction with tissue damage reminiscent of CD. In view of the potent autophagy induction in *Xbp1* defective mice at the bottom of the crypts, PCs are likely to be the culprit of the CD-like phenotype of these mice. Indeed, Paneth-specific ablation of *Xbp1* and *Atg16L1* (or *Atg7*) recapitulated the phenotype of the above mice where the same genes were inhibited in the whole intestinal epithelium, thus providing additional experimental evidence for PCs as the cellular site where ER stress and autophagy are functionally linked and as the origin of intestinal inflammation in IBD cases linked to ATG and ER stress loci (Adolph et al., 2013). As for the molecular mechanisms underlying the autophagy-driven amelioration of ER stress, upregulation of the NF-κB pathway was observed in the double-mutant mice. Accordingly, use of an NF-κB inhibitor led to a decrease in the overall number of cells undergoing apoptosis and a partial reversion of the inflammatory phenotype caused by the combined *Xbp1* and *Atg16L1* genetic deletion. Notably, NF-κB activation upon concomitant loss of *Xbp1* and *Atg16L1* function was not observed in germ-free mice again highlighting the relevance of the microbiome in these experimental settings and providing a suggestive functional connection between ER stress, autophagy, and the microbiome in IBD (Adolph et al., 2013).

The above findings open novel therapeutic scenarios for IBD patients, as illustrated by the use of NF-κB inhibitors to ameliorate and possibly resolve inflammation. Moreover, patients without SNPs in *Atg*-related genes, autophagy could be stimulated by treatment with rapamycin and mimicking caloric restriction (Yilmaz et al., 2012). Also, it is well established that IBD patients carry an increased risk to colon cancer. From this perspective, the notion that PCs are the site of origin of inflammation may also have implications for IBD-associated colon cancers in view of previous observations pointing at the capacity of PCs and their progenitors to dedifferentiate and take active part in the regenerative response to tissue damage (Roth et al., 2012). Further studies will be required to elucidate whether PCs might also represent the cell of origin of IBD-associated colorectal cancer.

3.4 The Role of Immunity and Environmental Factors

The immune system plays important roles in IBD in response to bacterial antigens, proteins, and RNA. Innate or in-born immunity is a prerequisite for the activation of the adaptive immunity which in turn seems to be causative for the tissue damage in IBD. Animal models of IBD, along with a more detailed understanding of the immune response in the digestive tract, have led to an unifying hypothesis relative to the role of the immune system in IBD. An inappropriate mucosal immune response to normal intestinal components leads to a local imbalance in cytokines resulting in a neutrophil and monocyte influx with subsequent secretion of oxygen radicals and enzymes, leading to tissue damage (Brown & Mayer, 2007).

The incomplete penetrance of IBD animal models and the heterogeneity of the disease can be accounted for by additional environmental factors. Several epidemiological studies have shown increased IBD risk in migrants from countries at low- to high-risk countries (Segal, 2016). Furthermore, since the 19th century the incidence of IBD has steadily increased in westernized countries (Segal, 2016). Diet and vitamin D intake are supposedly additional modifying risk factors to IBD. Low dietary fiber intake is associated with increased IBD risk as fibers are metabolized by bacteria into short-chain fatty acids that inhibit transcription of proinflammatory mediators (Ananthakrishnan et al., 2013, 2014). Furthermore, low zinc intake might impair autophagy and modulate immune functions (Ananthakrishnan et al., 2015; Liuzzi, Guo, Yoo, & Stewart, 2014). Additional factors correlated with higher risk of developing IBD are lack of exercise and sleep, although more experimental evidence is needed to link the latter functionally to IBD (Blumberg, 2016).

3.5 PCs as IBD-Specific Disease Markers

The broad range of processes coming together in IBD is indicative of the complexity and heterogeneity of the disease (Figure 2A). In view of their striking multifunctionality ranging from the secretory and bactericidal function to innate gut immunity, PCs seem to be endowed with the capacity to integrate cues from dietary nutrients and the microbiota and translate them in to stem cell niche regulatory signals in homeostasis and tissue regeneration. Accordingly, changes in PCs numbers and/or morphology or mutations in PC-specific genes underlie the etiology of at least a subset of IBD cases (Stappenbeck & McGovern, 2017). Stappenbeck and McGovern have proposed a classification of CD subtypes based on the detailed characterization of PC morphology (granules size and numbers) and IHC staining of AMPs like lysozyme (Lyz1) and defensins (Stappenbeck & McGovern, 2017). In this fashion PCs can be studied retrospectively in archival resection specimens or from biopsies collected via endoscopy.

3.6 PCs as Stem Cell Niche in Inflammation

The stem cell niche role of PCs has been mostly studied in homeostasis, and little is known on their function upon inflammation. Several reports suggest a decrease in mature PCs' multiplicity in inflammation. One of the genes responsible for the incomplete maturation of PCs in inflammation is protein kinase C iota or (*Prcki*). *Prcki* ablation in the intestinal epithelium did not affect PC number and granules. However, the PC-specific marker lysozyme was shown to colocalize with alcian blue-positive cells characteristic of the goblet lineage. Hence, inflammation triggers PC dedifferentiation to acquire an intermediate secretory precursor of both Paneth and goblet cells (Nakanishi et al., 2016). Expression analysis of PRCKI in IBD patients revealed an inverse correlation with disease progression (Nakanishi et al., 2016).

The secretory phospholipase A2 group 2a (*Pla2g2a*) gene was shown to be able to modify and improve the PC niche function upon inflammation (Schewe et al., 2016). Contrary to the role of *Pla2g2a* in homeostasis where the protein downregulates Wnt signaling in PCs, gene- and protein-based assays showed highly increased *Pla2g2a* expression and secretion upon inflammation leading to an increase in canonical Wnt in PCs. When secreted, Pla2g2a, next to exerting its antibacterial function, binds to the M-type phospholipase A2 receptor 1 (Pla2R1). This interaction triggers an autocrine cascade of downstream events culminating in increased Wnt signaling and improved regenerative response to the inflammatory tissue damage (Schewe et al., 2016). These observations open new therapeutic options by stimulating PC functions in IBD to improve tissue regeneration (Schewe et al., 2016).

4. CONCLUSIVE REMARKS

4.1 Multitasking PCs in Health and Disease

The strikingly broad spectrum of functions covered by PCs is reflected by the plethora of molecular and cellular mechanisms through which these secretory cells have been linked to several human pathological conditions related to dysbiosis, from obesity to graft vs host disease, inflammation, and cancer. The more recently elucidated niche role of PCs in supporting and modulating intestinal stem cell function possibly represents the integration hub of their multiple tasks. By sensing environmental factors like dietary changes and/or inflammatory stimuli, PCs seem to be capable of translating them into modifications of stem cell numbers, activity, and regenerative capacity. This is achieved either directly or indirectly by coordinated stress sensors such as UPR, ER stress, autophagy, and possibly many others. And yet, we are possibly only scratching the surface of the true functional identity of PCs. Many fascinating aspects remain unsolved: what is, if any, the functional role of PCM in colonic inflammation and cancer? Can mature PCs and their secretory precursors, next to their niche function, directly contribute to the regenerative response to tissue insults by dedifferentiating and acquiring stem-like properties? Analogous to their regulatory role in response to caloric restrictions, can Paneth-like cells located in the colon alter their metabolic needs and niche functions upon Western-style nutrients known to increase colon cancer risk? Ongoing and future studies will contribute to the elucidation of these and other questions and possibly lead to PC-tailored therapeutic strategies for IBD and GI-tract cancer patients (Figs. 1 and 2).

REFERENCES

Adolph, T. E., et al. (2013). Paneth cells as a site of origin for intestinal inflammation. *Nature*, *503*, 272–276. https://doi.org/10.1038/nature12599.

Akcora, D., et al. (2013). The CSF-1 receptor fashions the intestinal stem cell niche. *Stem Cell Research*, *10*, 203–212. https://doi.org/10.1016/j.scr.2012.12.001.

Amid, C., et al. (2009). Manual annotation and analysis of the defensin gene cluster in the C57BL/6J mouse reference genome. *BMC Genomics*, *10*, 606. https://doi.org/10.1186/1471-2164-10-606.

Ananthakrishnan, A. N., et al. (2013). A prospective study of long-term intake of dietary fiber and risk of Crohn's disease and ulcerative colitis. *Gastroenterology*, *145*, 970–977. doi: S0016-5085(13)01140-2 [pii] https://doi.org/10.1053/j.gastro.2013.07.050.

Ananthakrishnan, A. N., et al. (2014). Long-term intake of dietary fat and risk of ulcerative colitis and Crohn's disease. *Gut*, *63*, 776–784. doi:gutjnl-2013-305304 [pii] https://doi.org/10.1136/gutjnl-2013-305304.

Ananthakrishnan, A. N., et al. (2015). Zinc intake and risk of Crohn's disease and ulcerative colitis: A prospective cohort study. *International Journal of Epidemiology*, *44*, 1995–2005. doi:dyv301 [pii] https://doi.org/10.1093/ije/dyv301.

Andersson-Rolf, A., et al. (2016). Simultaneous paralogue knockout using a CRISPR-concatemer in mouse small intestinal organoids. *Developmental Biology*, *420*, 271–277. https://doi.org/10.1016/j.ydbio.2016.10.016.

Andersson-Rolf, A., et al. (2017). One-step generation of conditional and reversible gene knockouts. *Nature Methods, 14*, 287–289. https://doi.org/10.1038/nmeth.4156.
Ayabe, T., et al. (2000). Secretion of microbicidal alpha-defensins by intestinal Paneth cells in response to bacteria. *Nature Immunology, 1*, 113–118. https://doi.org/10.1038/77783.
Ayabe, T., et al. (2002). Modulation of mouse Paneth cell alpha-defensin secretion by mIKCa1, a Ca2+-activated, intermediate conductance potassium channel. *The Journal of Biological Chemistry, 277*, 3793–3800. https://doi.org/10.1074/jbc.M107507200.
Barker, N., et al. (2007). Identification of stem cells in small intestine and colon by marker gene Lgr5. *Nature, 449*, 1003–1007.
Barriga, F. M., et al. (2017). Mex3a marks a slowly dividing subpopulation of Lgr5+ intestinal stem cells. *Cell Stem Cell, 20*(6), 801–816. https://doi.org/10.1016/j.stem.2017.02.007.
Bastide, P., et al. (2007). Sox9 regulates cell proliferation and is required for Paneth cell differentiation in the intestinal epithelium. *The Journal of Cell Biology, 178*, 635–648. https://doi.org/10.1083/jcb.200704152.
Batlle, E., et al. (2002). Beta-catenin and TCF mediate cell positioning in the intestinal epithelium by controlling the expression of EphB/ephrinB. *Cell, 111*, 251–263.
Bevins, C. L. (2013). Innate immune functions of alpha-defensins in the small intestine. *Digestive Diseases, 31*, 299–304. https://doi.org/10.1159/000354681.
Bevins, C. L., & Salzman, N. H. (2011). The potter's wheel: The host's role in sculpting its microbiota. *Cellular and Molecular Life Sciences, 68*, 3675–3685.
Bizzozero, G. (1892). Ueber die schlauchförmigen Druisen des Magendarmkanals und die Beziehungen ihres Epithels zu dem Oberflachenepithel der Schleimhaut. *Archiv für Mikroskopische Anatomie, 40*, 325–375.
Blumberg, R. S. (2016). Environment and genes: What is the interaction? *Digestive Diseases, 34*, 20–26. https://doi.org/10.1159/000442920.
Brown, S. J., & Mayer, L. (2007). The immune response in inflammatory bowel disease. *The American Journal of Gastroenterology, 102*, 2058–2069. https://doi.org/10.1111/j.1572-0241.2007.01343.x.
Cadwell, K., Patel, K. K., Komatsu, M., Virgin, H. W., & Stappenbeck, T. S. (2009). A common role for Atg16L1, Atg5 and Atg7 in small intestinal Paneth cells and Crohn disease. *Autophagy, 5*, 250–252.
Cadwell, K., et al. (2008). A key role for autophagy and the autophagy gene Atg16l1 in mouse and human intestinal Paneth cells. *Nature, 456*, 259–263. https://doi.org/10.1038/nature07416.
Candela, M., et al. (2011). Human intestinal microbiota: Cross-talk with the host and its potential role in colorectal cancer. *Critical Reviews in Microbiology, 37*, 1–14. https://doi.org/10.3109/1040841X.2010.501760.
Carlone, D. L., & Breault, D. T. (2011). Slowly cycling versus rapidly cycling intestinal stem cells: Distinct roles or redundancy. *Cell Cycle, 10*, 723–724. https://doi.org/10.4161/cc.10.5.14867.
Chen, H. J., et al. (2016). A recellularized human colon model identifies cancer driver genes. *Nature Biotechnology, 34*, 845–851. https://doi.org/10.1038/nbt.3586.
Cheng, H. (1974). Origin, differentiation and renewal of the four main epithelial cell types in the mouse small intestine. IV. Paneth cells. *The American Journal of Anatomy, 141*, 521–535. https://doi.org/10.1002/aja.1001410406.
Cho, J. H. (2008). The genetics and immunopathogenesis of inflammatory bowel disease. *Nature Reviews. Immunology, 8*, 458–466.
Claesson, M. J., et al. (2012). Gut microbiota composition correlates with diet and health in the elderly. *Nature, 488*, 178–184.

Clarke, R. M. (1972). The effect of growth and of fasting on the number of villi and crypts in the small intestine of the albino rat. *Journal of Anatomy, 112*, 27–33.

Clevers, H. C., & Bevins, C. L. (2013). Paneth cells: Maestros of the small intestinal crypts. *Annual Review of Physiology, 75*, 289–311. https://doi.org/10.1146/annurev-physiol-030212-183744.

Cowdry, E. V. (1934). *A textbook of histology. Functional significance of cells and intercellular substances*. Lea and Febiger.

Creamer, B. (1967a). Paneth-cell function. *Lancet, 1*, 314–316.

Creamer, B. (1967b). The turnover of the epithelium of the small intestine. *British Medical Bulletin, 23*, 226–230.

Creamer, B., & Pink, I. J. (1967). Paneth-cell deficiency. *Lancet, 1*, 304–306.

Darmoul, D., Brown, D., Selsted, M. E., & Ouellette, A. J. (1997). Cryptdin gene expression in developing mouse small intestine. *The American Journal of Physiology, 272*, G197–206.

Das, S., et al. (2015). Rab8a vesicles regulate Wnt ligand delivery and Paneth cell maturation at the intestinal stem cell niche. *Development, 142*, 2147–2162. https://doi.org/10.1242/dev.121046.

Deckx, R. J., Vantrappen, G. R., & Parein, M. M. (1967). Localization of lysozyme activity in a Paneth cell granule fraction. *Biochimica et Biophysica Acta, 139*, 204–207.

Dietrich, W. F., et al. (1993). Genetic identification of Mom-1, a major modifier locus affecting Min-induced intestinal neoplasia in the mouse. *Cell, 75*, 631–639.

Dumas, M. E., et al. (2006). Metabolic profiling reveals a contribution of gut microbiota to fatty liver phenotype in insulin-resistant mice. *Proceedings of the National Academy of Sciences of the United States of America, 103*, 12511–12516. https://doi.org/10.1073/pnas.0601056103.

Durand, A., et al. (2012). Functional intestinal stem cells after Paneth cell ablation induced by the loss of transcription factor Math1 (Atoh1). *Proceedings of the National Academy of Sciences of the United States of America, 109*, 8965–8970. https://doi.org/10.1073/pnas.1201652109.

Farin, H. F., Van Es, J. H., & Clevers, H. (2012). Redundant sources of Wnt regulate intestinal stem cells and promote formation of Paneth cells. *Gastroenterology, 143*. 1518-1529.e1517. https://doi.org/10.1053/j.gastro.2012.08.031.

Farin, H. F., et al. (2014). Paneth cell extrusion and release of antimicrobial products is directly controlled by immune cell-derived IFN-gamma. *The Journal of Experimental Medicine, 211*, 1393–1405. https://doi.org/10.1084/jem.20130753.

Farin, H. F., et al. (2016). Visualization of a short-range Wnt gradient in the intestinal stem-cell niche. *Nature, 530*, 340–343. https://doi.org/10.1038/nature16937.

Feng, Y., et al. (2013). Sox9 induction, ectopic Paneth cells, and mitotic spindle axis defects in mouse colon adenomatous epithelium arising from conditional biallelic Apc inactivation. *The American Journal of Pathology, 183*, 493–503. https://doi.org/10.1016/j.ajpath.2013.04.013.

Fre, S., et al. (2005). Notch signals control the fate of immature progenitor cells in the intestine. *Nature, 435*, 964–968. https://doi.org/10.1038/nature03589.

Fuchs, E. (2009). The tortoise and the hair: Slow-cycling cells in the stem cell race. *Cell, 137*, 811–819. https://doi.org/10.1016/j.cell.2009.05.002.

Fujii, M., et al. (2016). A colorectal tumor organoid library demonstrates progressive loss of niche factor requirements during tumorigenesis. *Cell Stem Cell, 18*, 827–838. https://doi.org/10.1016/j.stem.2016.04.003.

Georges, M. (1930). Recherches cytologiques et experimentales sur les cellules de Paneth (souris). *Archives de Biologie, 40*, 111–150.

Glick, D., Barth, S., & Macleod, K. F. (2010). Autophagy: Cellular and molecular mechanisms. *The Journal of Pathology, 221*, 3–12. https://doi.org/10.1002/path.2697.

Gorbunov, N. V., & Kiang, J. G. (2009). Up-regulation of autophagy in small intestine Paneth cells in response to total-body gamma-irradiation. *The Journal of Pathology*, *219*, 242–252. https://doi.org/10.1002/path.2591.

Gracz, A. D., et al. (2015). A high-throughput platform for stem cell niche co-cultures and downstream gene expression analysis. *Nature Cell Biology*, *17*, 340–349. https://doi.org/10.1038/ncb3104.

Gregorieff, A., et al. (2009). The ets-domain transcription factor Spdef promotes maturation of goblet and paneth cells in the intestinal epithelium. *Gastroenterology*, *137*. 1333-1345. e1331-1333. https://doi.org/10.1053/j.gastro.2009.06.044.

Gulati, A. S., et al. (2012). Mouse background strain profoundly influences Paneth cell function and intestinal microbial composition. *PLoS One*, *7*, e32403. https://doi.org/10.1371/journal.pone.0032403.

He, C., & Klionsky, D. J. (2009). Regulation mechanisms and signaling pathways of autophagy. *Annual Review of Genetics*, *43*, 67–93. https://doi.org/10.1146/annurev-genet-102808-114910.

Hertzog, A. J. (1937). The Paneth cell. *The American Journal of Pathology*, *13*, 351–360. 351.

Hugot, J. P., et al. (2001). Association of NOD2 leucine-rich repeat variants with susceptibility to Crohn's disease. *Nature*, *411*, 599–603.

Huynh, D., et al. (2009). Colony stimulating factor-1 dependence of paneth cell development in the mouse small intestine. *Gastroenterology*, *137*, 136–144. 144.e131-133. https://doi.org/10.1053/j.gastro.2009.03.004.

Huynh, D., et al. (2013). CSF-1 receptor-dependent colon development, homeostasis and inflammatory stress response. *PLoS One*, *8*, e56951. https://doi.org/10.1371/journal.pone.0056951.

Igarashi, M., & Guarente, L. (2016). mTORC1 and SIRT1 cooperate to foster expansion of gut adult stem cells during calorie restriction. *Cell*, *166*, 436–450. doi:S0092-8674(16)30595-5 [pii]. https://doi.org/10.1016/j.cell.2016.05.044.

Jones, D. E., & Bevins, C. L. (1992). Paneth cells of the human small intestine express an antimicrobial peptide gene. *The Journal of Biological Chemistry*, *267*, 23216–23225.

Jones, D. E., & Bevins, C. L. (1993). Defensin-6 mRNA in human Paneth cells: Implications for antimicrobial peptides in host defense of the human bowel. *FEBS Letters*, *315*, 187–192.

Kabiri, Z., et al. (2014). Stroma provides an intestinal stem cell niche in the absence of epithelial Wnts. *Development*, *141*, 2206–2215. https://doi.org/10.1242/dev.104976.

Karlsson, F. H., et al. (2012). Symptomatic atherosclerosis is associated with an altered gut metagenome. *Nature Communications*, *3*, 1245. https://doi.org/10.1038/ncomms2266.

Kaser, A., Zeissig, S., & Blumberg, R. S. (2010). Inflammatory bowel disease. *Annual Review of Immunology*, *28*, 573–621. https://doi.org/10.1146/annurev-immunol-030409-101225.

Kaser, A., et al. (2008). XBP1 links ER stress to intestinal inflammation and confers genetic risk for human inflammatory bowel disease. *Cell*, *134*, 743–756. https://doi.org/10.1016/j.cell.2008.07.021.

Klein, S. (1906). On the nature of the granule cells of Paneth in the intestinal glands of mammals. *The American Journal of Anatomy*, *5*, 315–330.

Kobayashi, K. S., et al. (2005). Nod2-dependent regulation of innate and adaptive immunity in the intestinal tract. *Science*, *307*, 731–734. https://doi.org/10.1126/science.1104911.

Koo, B. K., et al. (2011). Controlled gene expression in primary Lgr5 organoid cultures. *Nature Methods*, *9*, 81–83. https://doi.org/10.1038/nmeth.1802.

Koren, O., et al. (2011). Human oral, gut, and plaque microbiota in patients with atherosclerosis. *Proceedings of the National Academy of Sciences of the United States of America*, *108*(Suppl. 1), 4592–4598. https://doi.org/10.1073/pnas.1011383107.

Langlands, A. J., et al. (2016). Paneth cell-rich regions separated by a cluster of Lgr5 + cells initiate crypt fission in the intestinal stem cell niche. *PLoS Biology, 14,* e1002491. https://doi.org/10.1371/journal.pbio.1002491.
Lechuga, S., Naydenov, N. G., Feygin, A., Jimenez, A. J., & Ivanov, A. I. (2017). A vesicle trafficking protein alphaSNAP regulates Paneth cell differentiation in vivo. *Biochemical and Biophysical Research Communications, 486,* 951–957. https://doi.org/10.1016/j.bbrc.2017.03.135.
Lehrer, R. I., & Lu, W. (2012). Alpha-Defensins in human innate immunity. *Immunological Reviews, 245,* 84–112. https://doi.org/10.1111/j.1600-065X.2011.01082.x.
Lewin, K. (1968). Neoplastic Paneth cells. *Journal of Clinical Pathology, 21,* 476–479.
Lewin, K. (1969a). The Paneth cell in health and disease. *Annals of the Royal College of Surgeons of England, 44,* 23–37.
Lewin, K. (1969b). The Paneth cell in disease. *Gut, 10,* 804–811.
Lewin, K. (1969c). Histochemical observations on Paneth cells. *Journal of Anatomy, 105,* 171–176.
Liu, T. C., et al. (2017). LRRK2 but not ATG16L1 is associated with Paneth cell defect in Japanese Crohn's disease patients. *JCI Insight, 2,* e91917. https://doi.org/10.1172/jci.insight.91917.
Liuzzi, J. P., Guo, L., Yoo, C., & Stewart, T. S. (2014). Zinc and autophagy. *Biometals, 27,* 1087–1096. https://doi.org/10.1007/s10534-014-9773-0.
MacPhee, M., et al. (1995). The secretory phospholipase A2 gene is a candidate for the Mom1 locus, a major modifier of ApcMin-induced intestinal neoplasia. *Cell, 81,* 957–966.
Mizushima, N., Yoshimori, T., & Ohsumi, Y. (2011). The role of Atg proteins in autophagosome formation. *Annual Review of Cell and Developmental Biology, 27,* 107–132. https://doi.org/10.1146/annurev-cellbio-092910-154005.
Möller, W. (1899). Anatomische Beitrage zur Frage von der Sekretion und Resorption in der Darmschleimhaut. *Zeitschrift für Wissenschaftliche Zoologie, 66,* 69–135.
Mori-Akiyama, Y., et al. (2007). SOX9 is required for the differentiation of paneth cells in the intestinal epithelium. *Gastroenterology, 133,* 539–546. https://doi.org/10.1053/j.gastro.2007.05.020.
Munoz, J., et al. (2012). The Lgr5 intestinal stem cell signature: Robust expression of proposed quiescent '+4' cell markers. *The EMBO Journal, 31,* 3079–3091. https://doi.org/10.1038/emboj.2012.166.
Mustata, R. C., et al. (2011). Lgr4 is required for Paneth cell differentiation and maintenance of intestinal stem cells ex vivo. *EMBO Reports, 12,* 558–564. https://doi.org/10.1038/embor.2011.52.
Nakanishi, Y., et al. (2016). Control of Paneth cell fate, intestinal inflammation, and tumorigenesis by PKClambda/iota. *Cell Reports, 16,* 3297–3310. https://doi.org/10.1016/j.celrep.2016.08.054.
Noah, T. K., Donahue, B., & Shroyer, N. F. (2011). Intestinal development and differentiation. *Experimental Cell Research, 317,* 2702–2710. https://doi.org/10.1016/j.yexcr.2011.09.006.
Ogura, Y., et al. (2001). A frameshift mutation in NOD2 associated with susceptibility to Crohn's disease. *Nature, 411,* 603–606.
Ogura, Y., et al. (2003). Expression of NOD2 in Paneth cells: A possible link to Crohn's ileitis. *Gut, 52,* 1591–1597.
Ootani, A., et al. (2009). Sustained in vitro intestinal epithelial culture within a Wnt-dependent stem cell niche. *Nature Medicine, 15,* 701–706. https://doi.org/10.1038/nm.1951.
Ouellette, A. J. (2011). Paneth cell alpha-defensins in enteric innate immunity. *Cellular and Molecular Life Sciences, 68,* 2215–2229. https://doi.org/10.1007/s00018-011-0714-6.

Ouellette, A. J., & Cordell, B. (1988). Accumulation of abundant messenger ribonucleic acids during postnatal development of mouse small intestine. *Gastroenterology*, *94*, 114–121.

Ouellette, A. J., Miller, S. I., Henschen, A. H., & Selsted, M. E. (1992). Purification and primary structure of murine cryptdin-1, a Paneth cell defensin. *FEBS Letters*, *304*, 146–148.

Paneth, J. (1888). Ueber die secernirenden Zellen des Dunndarm-Epithels. *Archiv für Mikroskopische Anatomie*, *31*, 113–191.

Paterson, J. C., & Watson, S. H. (1961). Paneth cell metaplasia in ulcerative colitis. *The American Journal of Pathology*, *38*, 243–249.

Potten, C. S., Kovacs, L., & Hamilton, E. (1974). Continuous labelling studies on mouse skin and intestine. *Cell and Tissue Kinetics*, *7*, 271–283.

Powell, R. H., & Behnke, M. S. (2017). WRN conditioned media is sufficient for in vitro propagation of intestinal organoids from large farm and small companion animals. *Biology Open*, *6*, 698–705. https://doi.org/10.1242/bio.021717.

Powell, A. E., et al. (2012). The pan-ErbB negative regulator Lrig1 is an intestinal stem cell marker that functions as a tumor suppressor. *Cell*, *149*, 146–158. https://doi.org/10.1016/j.cell.2012.02.042.

Qu, X. D., Lloyd, K. C., Walsh, J. H., & Lehrer, R. I. (1996). Secretion of type II phospholipase A2 and cryptdin by rat small intestinal Paneth cells. *Infection and Immunity*, *64*, 5161–5165.

Rodriguez-Colman, M. J., et al. (2017). Interplay between metabolic identities in the intestinal crypt supports stem cell function. *Nature*, *543*, 424–427. https://doi.org/10.1038/nature21673.

Ron, D., & Walter, P. (2007). Signal integration in the endoplasmic reticulum unfolded protein response. *Nature Reviews. Molecular Cell Biology*, *8*, 519–529. https://doi.org/10.1038/nrm2199.

Roth, S., et al. (2012). Paneth cells in intestinal homeostasis and tissue injury. *PLoS One*, *7*, e38965. https://doi.org/10.1371/journal.pone.0038965. PONE-D-12-10735 [pii].

Rumio, C., et al. (2004). Degranulation of paneth cells via toll-like receptor 9. *The American Journal of Pathology*, *165*, 373–381. https://doi.org/10.1016/S0002-9440(10)63304-4.

Rumio, C., et al. (2012). Induction of Paneth cell degranulation by orally administered Toll-like receptor ligands. *Journal of Cellular Physiology*, *227*, 1107–1113. https://doi.org/10.1002/jcp.22830.

Salzman, N. H., & Bevins, C. L. (2013). Dysbiosis—A consequence of Paneth cell dysfunction. *Seminars in Immunology*, *25*, 334–341. https://doi.org/10.1016/j.smim.2013.09.006.

Sangiorgi, E., & Capecchi, M. R. (2008). Bmi1 is expressed in vivo in intestinal stem cells. *Nature Genetics*, *40*, 915–920. https://doi.org/10.1038/ng.165.

Sato, T., et al. (2009). Single Lgr5 stem cells build crypt-villus structures in vitro without a mesenchymal niche. *Nature*, *459*, 262–265.

Sato, T., et al. (2011). Paneth cells constitute the niche for Lgr5 stem cells in intestinal crypts. *Nature*, *469*, 415–418.

Satoh, Y. (1988). Atropine inhibits the degranulation of Paneth cells in ex-germ-free mice. *Cell and Tissue Research*, *253*, 397–402.

Satoh, Y., Habara, Y., Ono, K., & Kanno, T. (1995). Carbamylcholine- and catecholamine-induced intracellular calcium dynamics of epithelial cells in mouse ileal crypts. *Gastroenterology*, *108*, 1345–1356.

Schewe, M., et al. (2016). Secreted phospholipases A2 are intestinal stem cell niche factors with distinct roles in homeostasis, inflammation, and cancer. *Cell Stem Cell*, *19*, 38–51.

Schroder, M., & Kaufman, R. J. (2005). The mammalian unfolded protein response. *Annual Review of Biochemistry*, *74*, 739–789. https://doi.org/10.1146/annurev.biochem.73.011303.074134.

Schwalbe, G. (1872). Beitriage zur Kenntniss der Driisen in den Darmwandungen, in's Besondere der Brunner'schen Driisen. *Archiv für Mikroskopische Anatomie*, *8*, 92–144.
Segal, A. W. (2016). Making sense of the cause of Crohn's—A new look at an old disease. *F1000Research*, *5*, 2510. https://doi.org/10.12688/f1000research.9699.2.
Senegas-Balas, F., et al. (1984). Immunohistochemical localization of intestinal phospholipase A2 in rat paneth cells. *Histochemistry*, *81*, 581–584.
Shanahan, M. T., Tanabe, H., & Ouellette, A. J. (2011). Strain-specific polymorphisms in Paneth cell alpha-defensins of C57BL/6 mice and evidence of vestigial myeloid alpha-defensin pseudogenes. *Infection and Immunity*, *79*, 459–473. https://doi.org/10.1128/IAI.00996-10.
Shroyer, N. F., Wallis, D., Venken, K. J., Bellen, H. J., & Zoghbi, H. Y. (2005). Gfi1 functions downstream of Math1 to control intestinal secretory cell subtype allocation and differentiation. *Genes & Development*, *19*, 2412–2417. https://doi.org/10.1101/gad.1353905.
Shroyer, N. F., et al. (2007). Intestine-specific ablation of mouse atonal homolog 1 (Math1) reveals a role in cellular homeostasis. *Gastroenterology*, *132*, 2478–2488. https://doi.org/10.1053/j.gastro.2007.03.047.
Snippert, H. J., et al. (2010). Intestinal crypt homeostasis results from neutral competition between symmetrically dividing Lgr5 stem cells. *Cell*, *143*, 134–144. https://doi.org/10.1016/j.cell.2010.09.016.
Stamos, J. L., & Weis, W. I. (2013). The beta-catenin destruction complex. *Cold Spring Harbor Perspectives in Biology*, *5*, a007898.
Stappenbeck, T. S., & McGovern, D. P. (2017). Paneth cell alterations in the development and phenotype of Crohn's disease. *Gastroenterology*, *152*, 322–326. https://doi.org/10.1053/j.gastro.2016.10.003.
St Clair, W. H., & Osborne, J. W. (1985). Crypt fission and crypt number in the small and large bowel of postnatal rats. *Cell and Tissue Kinetics*, *18*, 255–262.
Symonds, D. A. (1974). Paneth cell metaplasia in diseases of the colon and rectum. *Archives of Pathology*, *97*, 343–347.
Takeda, N., et al. (2011). Interconversion between intestinal stem cell populations in distinct niches. *Science (New York, NY)*, *334*, 1420–1424. https://doi.org/10.1126/science.1213214.
Tetteh, P. W., et al. (2016). Replacement of lost Lgr5-positive stem cells through plasticity of their enterocyte-lineage daughters. *Cell Stem Cell*, *18*, 203–213. https://doi.org/10.1016/j.stem.2016.01.001.
Tian, H., et al. (2015). Opposing activities of Notch and Wnt signaling regulate intestinal stem cells and gut homeostasis. *Cell Reports*, *11*, 33–42. https://doi.org/10.1016/j.celrep.2015.03.007.
Trier, J. S. (1966). The Paneth cells: An enigma. *Gastroenterology*, *51*, 560–562.
Trier, J. S. (1967). Structure of the mucosa of the small intestine as it relates to intestinal function. *Federation Proceedings*, *26*, 1391–1404.
Turnbaugh, P. J., et al. (2006). An obesity-associated gut microbiome with increased capacity for energy harvest. *Nature*, *444*, 1027–1031. https://doi.org/10.1038/nature05414.
Vaishnava, S., Behrendt, C. L., Ismail, A. S., Eckmann, L., & Hooper, L. V. (2008). Paneth cells directly sense gut commensals and maintain homeostasis at the intestinal host-microbial interface. *Proceedings of the National Academy of Sciences of the United States of America*, *105*, 20858–20863. https://doi.org/10.1073/pnas.0808723105.
van de Wetering, M., et al. (2002). The beta-catenin/TCF-4 complex imposes a crypt progenitor phenotype on colorectal cancer cells. *Cell*, *111*, 241–250.
VanDussen, K. L., et al. (2012). Notch signaling modulates proliferation and differentiation of intestinal crypt base columnar stem cells. *Development*, *139*, 488–497. https://doi.org/10.1242/dev.070763.

VanDussen, K. L., et al. (2014). Genetic variants synthesize to produce paneth cell phenotypes that define subtypes of Crohn's disease. *Gastroenterology*, *146*, 200–209. https://doi.org/10.1053/j.gastro.2013.09.048.
van Es, J. H., et al. (2005). Wnt signalling induces maturation of Paneth cells in intestinal crypts. *Nature Cell Biology*, *7*, 381–386.
Vermeulen, L., et al. (2010). Wnt activity defines colon cancer stem cells and is regulated by the microenvironment. *Nature Cell Biology*, *12*, 468–476. https://doi.org/10.1038/ncb2048.
Walter, P., & Ron, D. (2011). The unfolded protein response: From stress pathway to homeostatic regulation. *Science (New York, NY)*, *334*, 1081–1086. https://doi.org/10.1126/science.1209038.
Wu, G. D., et al. (2011). Linking long-term dietary patterns with gut microbial enterotypes. *Science (New York, NY)*, *334*, 105–108.
Yan, K. S., et al. (2012). The intestinal stem cell markers Bmi1 and Lgr5 identify two functionally distinct populations. *Proceedings of the National Academy of Sciences of the United States of America*, *109*, 466–471. https://doi.org/10.1073/pnas.1118857109.
Yang, Q., Bermingham, N. A., Finegold, M. J., & Zoghbi, H. Y. (2001). Requirement of Math1 for secretory cell lineage commitment in the mouse intestine. *Science (New York, NY)*, *294*, 2155–2158. https://doi.org/10.1126/science.1065718.
Yilmaz, O. H., et al. (2012). mTORC1 in the Paneth cell niche couples intestinal stem-cell function to calorie intake. *Nature*, *486*, 490–495. https://doi.org/10.1038/nature11163.
Zhang, Q., et al. (2015). Commensal bacteria direct selective cargo sorting to promote symbiosis. *Nature Immunology*, *16*, 918–926. https://doi.org/10.1038/ni.3233.
Zorn, A. M., & Wells, J. M. (2009). Vertebrate endoderm development and organ formation. *Annual Review of Cell and Developmental Biology*, *25*, 221–251. https://doi.org/10.1146/annurev.cellbio.042308.113344.

CHAPTER THREE

Intestinal Stem Cells and Their Niche at Homeostasis and Under Stress

Lucie Peduto[*,†], **Philippe J. Sansonetti**[‡,§,¶,1], **Giulia Nigro**[‡,§,2]
[*]Unité Stroma, Inflammation & Tissue Repair, Institut Pasteur, Paris, France
[†]INSERM U1224, Paris, France
[‡]Unité de Pathogénie Microbienne Moléculaire, Institut Pasteur, Paris, France
[§]INSERM U1202, Paris, France
[¶]Chaire de Microbiologie et Maladies Infectieuses, Collège de France, Paris, France
[1]Corresponding author: e-mail address: philippe.sansonetti@pasteur.fr

Contents

1. The Gut: From Structure to Regeneration 77
   1.1 Epithelial Layer 78
   1.2 The Lamina Propria 80
2. Concept of Niche 83
   2.1 Paneth Cells 84
   2.2 Mesenchymal Cells 86
   2.3 Microbiota 88
3. Conclusions 92
References 93

1. THE GUT: FROM STRUCTURE TO REGENERATION

The intestinal mucosa is composed of a continuous single layer of epithelial cells bound to a basement membrane overlying a layer of connective tissue, the lamina propria, containing mesenchymal cells such as subepithelial myofibroblasts, fibroblasts, pericytes, and immune cells, and smooth muscle cells surrounding the lymphatic lacteals. The lamina propria is separated from the submucosa by a thin layer of smooth muscle cells, the muscularis mucosae.

[2] Senior author.

Advances in Stem Cells and their Niches, Volume 2
ISSN 2468-5097
https://doi.org/10.1016/bs.asn.2017.12.002

1.1 Epithelial Layer

The intestinal epithelium is organized in structures with protrusions, the villi, and invaginations, the crypts. These two regions have different functions: mainly the villi are designated to the absorption, while the crypts are the regeneration apparatus, organized around the intestinal stem cells (ISCs) that are located at the bottom of the crypts.

The gut epithelium continuously copes with hazards linked to its digestive and barrier functions. The gut microbiota, including both pathogens and commensal bacteria, and compounds present in the food, such as xenobiotics (drugs, food additives, and environmental pollutants) and carcinogens, influence the integrity and the physiology of the gut epithelium. The epithelial cells that are continuously exposed to these insults present a high turnover, and while dying cells are shed from the top of the villi, new cells are generated from resident adult stem cells located in the stem cells zone at the bottom. Due to this continuous renewal process, the intestinal epithelium is one of the most rapidly self-renewing tissues in mammals. In homeostatic conditions, a tight equilibrium is achieved between the rate of cell death and the rate of generation of new differentiated cells. ISCs generate highly proliferative cells, forming the transit-amplifying (TA) compartment. These cells, upon four to five divisions, move out of the crypts, differentiating in five of the six different cell types composing the intestinal epithelium. The sixth type, the Paneth cells, migrates downward in the crypt.

Although the existence of self-renewing, multipotent ISCs in the crypt is well described in the literature, the exact location and identity of these cells are still controversial. Two models are described: the +4 model and the stem cell zone model.

The *+4 model*, principally described by Potten, Hume, Reid, and Cairns (1978), suggested that ISCs are located just above the Paneth cells, approximately four cells from the base of the crypt. This cell population was identified using long-term DNA labeling with ^{3}H-thymidine and bromodeoxyuridine (Potten, Owen, & Booth, 2002) showing that cells in the +4 position present DNA label retention over several weeks, revealing mitotic quiescence, a characteristic that, at the time, was considered a marker of stemness. For several decades, this remained the major model to describe ISCs. The alternative *stem cell zone model* was initially proposed by Cheng and Leblond (1974a). Using electron microscopy, they discovered slender, immature proliferative cells intercalated between Paneth cells, called crypt-base columnar cells (CBCs). Using ^{3}H-thymidine and "phagosome" tracing, they concluded that CBCs were able to differentiate in the different intestinal epithelial cells, therefore behaving as stem cells (Cheng & Leblond, 1974b). The ISC field remained "quiet" for several years, mainly due to the lack of functional stem cells markers and in vitro assays showing the stemness capacity of a specific population. Only recently, this field has undergone a major advance when the Hans Clevers' group identified the Wnt target gene *Lgr5* as a specific marker for the CBC population (Barker et al., 2007). They generated an Lgr5 reporter mouse line that allowed them to carry out in vivo lineage tracing, showing that all the intestinal epithelium was generated from CBC-Lgr5-expressing cells. Moreover, they were able to establish in vitro growth conditions to validate the stem cell potential of these cells: isolated Lgr5+ cells were able to generate in vitro tridimensional structures, defined as organoids, mimicking minigut with a lumen, crypt-like structures, and containing all the epithelial lineages, including self-autonomous ISCs (Sato et al., 2009).

Several studies have tried to identify markers for the slow-cycling/quiescent +4 ISCs (mTert, Lrig1, and Hopx) (Smith, Gallagher, & Wong, 2016). The Capecchi's group identified Bmi1 as possible main marker of this population (Sangiorgi & Capecchi, 2008). Using a Bmi1 reporter mouse, they showed that Bmi1-expressing cells were slowly cycling at homeostasis, but were able to expand in response to injury achieving regeneration of the tissue. It has been proposed that this cell population functions as reserve for the fast-cycling ISCs upon their depletion. Despite this initial report, other studies have documented the expression of Lgr5 in putative +4 cells (Muñoz et al., 2012), including the Bmi1+ cells in the pool of Lgr5 cells.

However, crypts display a remarkable regenerative capacity following stress, DNA or cytotoxic damage, and injuries. As alternative to the quiescent pool of ISCs, it has been proposed that cycling secretory progenitors Dll1+ cells (van Es et al., 2012) as well as quiescent secretory progenitors (label-retaining cells) (Buczacki et al., 2013) could dedifferentiate to ISCs in vivo to replenish a loss of ISCs. Recently, another pool of progenitors was proposed as possible reservoir of ISC: these cells are enterocyte precursors, expressing intestinal alkaline phosphate (Alpi) (Tetteh et al., 2016). The different models of damage described could explain the intervention of one of these pools of progenitors, but this remains to be validated, and it remains unclear if two different populations really exist, i.e., a slow-cycling/quiescent ISC at position 4, and fast-cycling ISCs in the stem cells zone, or if they just represent the same population (Clevers, 2013; Smith et al., 2016). Despite the specific signatures of these two populations, the general model suggests the presence of active cycling cells, and a "reserve" of ISCs, quiescent ISCs, or cells of the TA compartment, able to generate new ISCs (Smith et al., 2016).

Among the different cell populations in the intestinal epithelium, the most abundant cells are *enterocytes* that ensure the main respective functions of digestion/absorption of food/nutrients and water/electrolytes transport from the lumen. *Goblet cells* are the most abundant secretory lineage of the intestine and they are dispersed between villi and crypts. Their main function is to produce and secrete mucus creating a protective layer over the epithelium against the luminal content (Birchenough, Johansson, Gustafsson, Bergström, & Hansson, 2015). *Paneth cells* are the only terminally differentiated cell type present in the crypt. These cells participate in innate immunity, by secreting antimicrobial peptides and lysozyme, and as niche for stem cells, producing factors needed for the stemness maintenance (Clevers & Bevins, 2013). *Microfold (M)* cells are mainly found on the follicle-associated epithelium that overlies Peyer's patches in the small intestine and over lymphoid follicles in colon and rectum. They play a role in the development of host mucosal immunity uptaking luminal antigens and delivering them to the underlying immune cells (Miller, Zhang, Kuolee, Patel, & Chen, 2007). *Enteroendocrine cells* are scattered throughout the epithelium and produce gut hormones in response to luminal stimuli and thereby exert actions ranging from the local control of gut motility and secretion to the regulation of insulin release and food intake (Gribble & Reimann, 2017). *Tuft cells* are a rare population of secretory cells *distributed along villi and crypts*. Only very recently, three independent research articles have described the role of tuft cells in initiation of immune type 2 responses (Gerbe & Jay, 2016).

1.2 The Lamina Propria

The intestinal lamina propria refers to the layer of connective tissue lying beneath the epithelium and separated from the submucosa by the muscularis mucosae. The intestinal lamina propria contains blood vessels, lymphatics including the intestinal lacteals of the villi that absorb the dietary fat, immune cells, and organized lymphoid structures (cryptopatches, ILFs), as well as mesenchymal stromal cells.

The intestinal lamina propria contains the largest reservoir of immune cells besides lymphoid organs, which are essential to establish immune tolerance toward microbiota symbionts, and provide immune defense against potential pathogens. Surveillance of intestinal pathogens is ensured by a complex system including dendritic cells (DCs) extending dendrites into the lumen, epithelial cells equipped with pattern recognition receptors (PRRs) that induce chemokines to recruit leucocytes, intraepithelial lymphocytes, lamina propria DCs, macrophages, neutrophils, mast cells specialized in detecting pathogens, as well as lamina propria lymphocytes, and organized tertiary lymphoid organs (Iwasaki & Medzhitov, 2015). How these different types of immune cells develop and are maintained in the intestine, in constant bidirectional communication with the microbiota, has been the focus of many excellent reviews (Caballero & Pamer, 2015; Molloy, Bouladoux, & Belkaid, 2012) and will not be reviewed here.

1.2.1 Myofibroblastic Mesenchymal Cells

Besides immune cells, the lamina propria contains heterogeneous populations of mesenchymal stromal cells that constitute a 3D network necessary for the structure, organization, and functionality of the lamina propria. Intestinal mesenchymal cells are nonhematopoietic (CD45–) and nonendothelial (CD31–) cells and include subepithelial myofibroblasts, as well as mural cells embedded in the wall of blood vessels (such as pericytes and adventitial cells), smooth muscle cells associated with lymphatic lacteals or associated with arteries/veins, and smooth muscle cells of the muscularis mucosae. All these cell types can be identified by expression of alpha-smooth muscle actin (αSMA), an isoform of actin specific to smooth muscle cells. However, expression of αSMA varies importantly between these subsets. Notably, smooth muscle cells associated with lacteals and arteries as well as smooth muscle cells of the muscularis mucosae and of the muscularis externa express prominent levels of αSMA, which makes difficult to detect intestinal subepithelial myofibroblasts and mural cells by histology, as these populations express significantly lower levels of αSMA. This is often a source of confusion between these different cell types. In addition to αSMA, subepithelial myofibroblasts express a number of extracellular matrix (ECM) proteins including collagen, and proteins involved in the regulation of the ECM such as proyl-4-hydroxylase, angiogenic factors, as well as growth factors essential for survival and differentiation of epithelial cells such as BMPs and FGFs (Madison et al., 2005; van Dop et al., 2009). Consistent with a close communication with epithelial cells, subepithelial myofibroblasts are localized below the epithelial layer, where they form a thin layer of tightly interconnected cells. As epithelial cells ascend the crypt, BMPs produced by subepithelial myofibroblasts restrain Wnt-induced epithelial proliferation while promoting their differentiation into secretory or absorptive epithelial cells (Haramis et al., 2004; He, Zhang, Wei-Gang, Tawfik, & Ross, 2004). Subepithelial myofibroblasts are therefore essential for intestinal homeostasis. Interestingly, pericytes wrapping around capillaries express also some levels of αSMA, and subepithelial myofibroblasts are closely adjacent to capillaries below the epithelial layer, suggesting that pericytes or pericytes-like cells might function as subepithelial myofibroblasts (Powell, Pinchuk, Saada, Chen, & Mifflin, 2011).

In addition to specific subsets of myofibroblasts present at homeostasis, such as subepithelial myofibroblasts in the gut, myofibroblasts develop massively in areas of injury. The myofibroblast is the major mesenchymal cell type found in healing wounds, as well as in inflammatory and fibrotic pathologies affecting diverse organs including the intestine. In chronic diseases, including inflammatory bowel diseases, overexpression of collagen by myofibroblasts as well as other ECM proteins and proinflammatory cytokines plays a central role in disease pathogenesis.

The origin of intestinal myofibroblasts, both at homeostasis and after injury, remains unclear. During development, the neural crest and the serosal mesothelium have been shown to generate specific subsets of intestinal mesenchymal cells, such as endoneurial fibroblasts and smooth muscle cells of the vasculature (Wilm, Ipenberg, Hastie, Burch, & Bader, 2005). In adult intestine at homeostasis, inducible fate mapping of Grem1 +-expressing cells identified mesenchymal progenitors appearing first at the isthmus between crypt and villi, which gradually gave rise to a majority of subepithelial myofibroblasts in the intestinal villi, as well as to pericytes (Worthley et al., 2015). However, it remains to be determined whether these cells are involved in the generation of pathological myofibroblasts developing in intestinal inflammation/injury.

1.2.2 Nonmyofibroblastic Mesenchymal Cells

To facilitate identification of nonmyofibroblastic mesenchymal components of the lamina propria, a number of other markers have been used. Staining with CD90 (Thy-1), Fsp1, vimentin, PDGFRa, PDGFRb, Sca1, or CD34 revealed the existence of different subsets of intestinal mesenchymal cells which have no detectable expression of αSMA and are therefore not assimilated to myofibroblasts. These nonmyofibroblastic mesenchymal populations, often coexpressing Fsp1, CD90, or PDGFRa, are variously called intestinal fibroblasts or mesenchymal stromal cells. Specific subsets were suggested to be related to mesenchymal stem cells (MSCs) based on expression of markers expressed by MSCs of the bone marrow such as CD44 and CD29, and their capacity to differentiate into mesodermal lineages such as adipocytes and osteocytes, at least as a bulk population (Walker, Brown, Riehl, Stenson, & Stappenbeck, 2010). A subset of MSC-like stromal cells expressing Ptgs2 were essential to respond to microbe-derived signals triggering MyD88 and potentially other innate immune-signaling pathways (Brown et al., 2007).

The role of intestinal mesenchymal cells in immunity is slowly recognized. In lymphoid organs, specific subsets of specialized mesenchymal cells (also called lymphoid stromal cells, LSCs) are essential to recruit immune cells and organize immune responses (Chang & Turley, 2015). Expression of gp38 (podoplanin), which is constitutively expressed by LSCs of the T-cell zone, was shown to identify a major fraction of intestinal mesenchymal cells (Peduto et al., 2009). Further analysis of gp38+ intestinal stromal cells showed that coexpression of gp38 and CD34 identifies nonmyofibroblastic mesenchymal cells around the crypts (gp38+ CD34+ αSMA–), localized in close proximity to Lgr5+ ISC, while gp38+ CD34– αSMA+, a population enriched in myofibroblasts, are mainly localized in the intestinal villi and below the colonocytes. Consistent with a role in intestinal immunity, Gp38+ mesenchymal cells produced chemokines and survival factors essential for recruitment and survival of immune cells in DSS-induced colitis (Stzepourginski et al., 2017).

2. CONCEPT OF NICHE

Stem cells are located in a microenvironment, the stem cell niche, where they reside, receiving stimuli that determine their self-renewing and the capacity to produce progeny cells (Schofield, 1978). Therefore, the niche is a physical region where the stem cells are located, where extrinsic signals provide stimuli to stem cell influencing their capacity to survive, proliferate, and differentiate (Ferraro, Celso, & Scadden, 2010). On this context, a well-described niche for the intestinal stem cells is provided by nearby epithelial cells, particularly *Paneth cells*, as well as the surrounding *mesenchymal cells*. More recently, also the *microbiota* present in the lumen has been described as a niche component, releasing bacterial components and/or metabolites, modulating the composition of the "milieu" to which the ISCs are exposed, providing them with different stimuli.

The role of the niche is emphasized in stress situations, such as damages induced by pathogens, chemicals, or any kind of injuries that alter tissue integrity, and the ISCs respond by organizing repair and regeneration of the tissue.

2.1 Paneth Cells

Paneth cells are the only differentiated cells that escape the villi upward flow as they migrate downward to the bottom of the crypt, where they persist for 1 month. The oldest Paneth cells reside at the very base of the crypt. In this position, Paneth cells are localized in close association with CBCs (Cheng & Leblond, 1974b; Sato et al., 2011), therefore being good candidate to represent a niche component. The observation that Lgr5 + cells generate intestinal organoids in vitro indicates that mesenchymal components are not required to maintain ISCs in vitro. Exogenous factors, such as Wnt agonist (R-spondin1), BMP inhibitors (Noggin), Wnt, and Notch ligands (Wnt3 and Jagged-1), are sufficient to allow single stem cells, embedded in matrigel, to form minigut structures (Sato et al., 2009). Gene-expression profile studies carried out in the Clevers' group on Paneth cells revealed that they produce large amounts of Notch, Wnt3, and EFGR ligands such as EGF, which are sufficient to maintain the organoid culture. Moreover, the plating of doublets of Lgr5 + cells and Paneth cells increased the yield of organoids over 10-fold vs single ISCs alone (Sato et al., 2011). These in vitro findings demonstrate that Paneth cells provide essential niche signals to ISCs.

Still controversial is the situation in vivo. In 1997 the Gordon's group examined the role of Paneth cells by lineage ablation in mice (Garabedian, Roberts, McNevin, & Gordon, 1997). In these mice, upon induction, Paneth cells expressed diphtheria toxin, causing their apoptosis. At the time, this observation was that the ablation did not produce detectable effect on the proliferation or differentiation program of the other lineages rejecting the hypothesis of Paneth cells being required to maintain ISCs. Instead, the Clevers' group revisited this and other two models for Paneth cells loss: Gfi1 −/− mice described to lack Paneth cells (Shroyer, Wallis, Venken, Bellen, & Zoghbi, 2005) and conditional deletion of Sox9 (Bastide et al., 2007; Mori Akiyama et al., 2007). They showed that genetic ablation of Paneth cells in the three models resulted in the concomitant loss of Lgr5 + cells, and that tissue regeneration was supported by the fact that not all Paneth cells and ISCs were ablated in these models, allowing surviving ISCs to regenerate the full tissue. Therefore, Paneth cells are indispensable for in vitro growth of the miniguts, whereas they seem dispensable for in vivo niche functions.

Having shown the role of Paneth cells as niche component for ISCs, their absence in the colon opened the questions of the possible existence of Paneth-like cells in this tissue, at least to compensate their functions. Clarke's group identified cKit + colonic crypt base secretory cells characterized by the expression of niche factors, such as Dll1, Dll4, and EGF, able to support colon organoid formations (Rothenberg et al., 2012). Recently, the Clevers' group has further investigated this population of cells, emphasizing that these cells were originally described 30 years ago by Altmann, who named them "deep crypt secretory (DCS) cells" (Altmann, 1983). They identified Reg4 as a marker for DCS cells (Sasaki et al., 2016). The use of transgenic mice expressing human diphtheria toxin (DT) receptor in the Reg4+ cells allowed them to ablate this cell population in vivo upon administration of DT, showing the loss of Lgr5+ cells, supporting the hypothesis of their role as niche cells. As for Paneth cells, also DCS cells support the in vitro growth of sorted ISCs increasing the yield of organoids by 8%. In contrast to Paneth cells, DCS cells do not express Wnt ligands, implying the production of these signals from other cells, such as mesenchymal cells.

Recently, Rodriguez and collaborators have investigated if Paneth cells may also support the metabolic state of the ISCs (Rodríguez-Colman et al., 2017). In their publication, they showed that ISC display high mitochondrial activity, with an oxidative metabolic program, while Paneth cells mainly affect glycolytic activities. Lactate produced by Paneth cells, as end product of the glycolysis, strongly enhances the ability of ISCs to establish organoids. This effect is due to the conversion of lactate into pyruvate in ISCs, fueling mitochondrial OXPHOS, leading to ROS signaling and activation of p38 and differentiation. In conclusion, they were able to show that Paneth cells provide not only growth factor signaling but also metabolic elements to ISCs. Moreover, they could also show the important role of ROS in the homeostasis of ISCs, rather than a stress signal.

The role of Paneth cells as niche component for ISCs is also emphasized in situation of stress, such as in response to nutritional deprivation. Sabatini's group first, and more recently Guarente's group, investigated the signals of Paneth cells on ISCs during calorie restriction (Igarashi & Guarente, 2016; Yilmaz et al., 2012). In both papers, it was observed that Paneth cells could sense the nutritional state in the gut and modulate ISC responses. Particularly, although following calorie restriction a shrinkage of villi is observed, the ISC/PC compartment expands, with altered coupling between ISCs and TA cells. Yilmaz and collaborators, using in vivo and in vitro models, showed that this effect was dependent on the inhibition of mTORC1 signaling in Paneth cells, and not in ISCs, causing the increase of Bst1 levels and consequently the secretion of cyclic ADP ribose (cADPR) acting then as a proliferation inducer (Yilmaz et al., 2012). In contrast to the first paper, Igarashi et al. observed that calorie restrictions had a direct effect, not only on Paneth cells but also on ISCs. The cADPR released by Paneth cells triggers SIRT1 activity in ISCs with consequent mTORC1 signaling activation in these cells, promoting protein synthesis and proliferation. The authors commented these discrepant results in the ISCs by hypothesizing differences in methodologies.

Paradoxically, also under high-fat diet, Yilmaz showed that the number of ISCs increased, but this time with a concomitant decrease in Paneth cells (Beyaz et al., 2016). This effect is mediated by fatty acid-induced activation of PPAR-δ in ISCs, with the consequent activation of Wnt/β-catenin pathway that reduces the niche dependency of ISCs.

Fodde's group identified secreted phospholipases A2 (sPLA2s) as ISC niche factors (Schewe et al., 2016), with distinct roles during homeostasis and inflammation. Different groups of sPLA2 are expressed in small intestine by the Paneth cells and in the colon by Paneth-like cells (cKit+). At homeostasis, sPLA2s remain intracellular inhibiting Wnt signaling through the activation of Yap1. Upon inflammation, sPLA2 expression increases and it is secreted into the lumen. Extracellular sPLA2 stimulates therefore ISC proliferation inducting Wnt signaling.

2.2 Mesenchymal Cells

A complex gradient of factors maintains ISC stemness and proliferation and supports enterocyte differentiation along the crypt–villus axis. The Wnt signaling is required to maintain the ISC niche (Kuhnert et al., 2004; Pinto, Gregorieff, Begthel, & Clevers, 2003). Accordingly, abrogation of Wnt signaling by adenoviral expression of the Wnt antagonist Dickkopf-1 (Dkk1) or disruption of the *Tcf7/2* gene, encoding for Tcf4 (transcription factor 4), is sufficient to induce loss of crypt proliferation (Korinek et al., 1998; Kuhnert et al., 2004). In the crypts, Paneth cells express factors that promote stem cells growth, including Wnt3a, Notch ligands (Dll4, Dll1), and EGF (Pinto et al., 2003; Sato et al., 2011; van Es et al., 2005), and therefore were suggested to represent the major niche for ISCs (Sato et al., 2011). However, by inducing a partial or complete loss of Paneth cells, a number of groups showed that absence of Paneth cells does not affect crypt homeostasis in vivo (Durand et al., 2012; Garabedian et al., 1997; Mori Akiyama et al., 2007; Shroyer et al., 2005). Among these models, genetic disruption of the transcription factor *Math1* (also named *Atoh1*; atonal bHLH transcription factor 1), which has an essential role in differentiation of secretory cells, leads to complete loss of Paneth cells and other secretory cells. Using

Math1-inducible-deficient mice, Durand et al. showed that complete loss of Paneth cells does not affect crypt proliferation or β-catenin signaling. However, survival of Math1 KO organoids required addition of exogenous Wnts, suggesting that in vivo, nonepithelial sources can compensate for the lack of epithelial Wnts (Durand et al., 2012).

These results indicate the existence of redundant or compensation mechanisms in the crypts. A number of other cell types are present in the vicinity of crypts, including cells of hematopoietic and mesenchymal origin. Factors produced by mesenchymal stromal cells have an essential role in the maintenance of ISCs, such as Wnt2b (Farin, van Es, & Clevers, 2012), R-spondin1 (Rspo1) (de Lau et al., 2012; Kim et al., 2005; Ootani et al., 2009), and the BMP antagonist Gremlin1 (Grem1) (Hsu, Economides, Wang, Eimon, & Harland, 1998; Kosinski et al., 2007). Consistent with a role for the intestinal stroma as niche, total intestinal tissues depleted from epithelial cells, as well as cultured mesenchymal fractions enriched in myofibroblasts, were sufficient to compensate from the lack of functional Wnt signaling in Paneth cells of *Porcn/VillinCre*^{ERT2} mice (Kabiri et al., 2014). Myofibroblasts isolated from postnatal intestine lead to a culture of cells that supported growth of ISCs in organoids (Lei et al., 2014). In addition to Wnts, intestinal mesenchymal cells after injury or infection were shown to produce factors such as angiopoietin-like 2 (Angptl2), which maintains the ISC niche by modulating levels of competing signals between BMPs and β-catenin (Horiguchi et al., 2017), IL-33, which favors differentiation into secretory cell lineage to promote repair and antimicrobial defense (Mahapatro et al., 2016), and the EGFR ligand amphiregulin, which coordinated with BMP antagonists to elicit ISC proliferation in ex vivo cultures (Seiler et al., 2015).

Consistent with a mesenchymal but nonmyofibroblastic niche, San Roman and colleagues further showed that conditional ablation of *Porcn* in Myh11 (myosin heavy chain 11)-expressing cells, essentially myofibroblasts and smooth muscle cells, in *Porcn/Myh11Cre*^{ERT2} mice does not disrupt cell proliferation or Wnt activity in crypts, showing that myofibroblastic populations are not essential in vivo to maintain ISCs (San Roman, Jayewickreme, Murtaugh, & Shivdasani, 2014). More recently, diphtheria toxin-mediated ablation of intestinal mesenchymal cells, suggested to be distinct from myofibroblasts and expressing Foxl1 (winged-helix transcription factor forkhead box l 1), induced a cessation of epithelial proliferation and Wnt signaling, indicating an essential role for Foxl1+ mesenchymal cells in maintaining the ISC niche (Aoki et al., 2016).

One of the major issues that prevented a more precise definition of the mesenchymal niche for ISCs is the lack of specific surface markers to allow positive selection of defined mesenchymal subsets. Even though expression of αSMA allows identification of myofibroblasts by histology, its intracellular expression prevents its use as a marker to isolate live cells by FACS. Stzepourginski et al. recently reported that coexpression of gp38 and CD34 by FACS, in the absence of hematopoietic (CD45) or endothelial (CD31) markers, identifies pericryptal αSMA– mesenchymal cells that express highest levels of niche factors Wnt2b, Rsp1, and Grem1, compared to lamina propria immune cells, endothelial cells, and other stromal cells, and are sufficient to maintain ISCs in organoids (Stzepourginski et al., 2017). In contrast, αSMA+ myofibroblasts of the villi are enriched in the gp38+ CD34– mesenchymal subset, which produced high levels of prodifferentiation factors such as BMP2, BMP7, and Ptch1, and did not support ISCs in organoids (Stzepourginski et al., 2017). After DSS-induced colitis, pericryptal CD34 + stromal cells upregulated, in addition to niche factors, a number of genes with essential roles in recruitment and survival of immune cells, which potentially have a role in reconstruction of the niche. Consistent with this hypothesis, Pull et al. showed that macrophages activated after injury are essential for the regenerative response after DSS-induced injury in a microbiota-dependent way. These data suggest that intestinal macrophages coordinate input from the microbiota and injured epithelial cells to adapt ISCs proliferation to injury (Pull, Doherty, Mills, Gordon, & Stappenbeck, 2005), potentially through a mechanism involving mesenchymal cells. Accordingly, regeneration after injury was dependent on MyD88-mediated signaling in intestinal mesenchymal cells and subsequent induction of Ptgs2 expression, an enzyme essential for localized prostaglandin production mediating intestinal epithelial proliferation, resolution of inflammation, and tissue repair (Manieri, Drylewicz, Miyoshi, & Stappenbeck, 2012). Further studies showed that other signals from immune subsets might be essential for the niche, in particular after injury. Using ex vivo organoid cultures, Lindemans et al. showed that IL-22 produced by innate lymphoid cells after injury is essential to induce STAT3-dependent proliferation and expansion of ISCs, independently of Paneth cells (Lindemans et al., 2015).

2.3 Microbiota

2.3.1 Microbiota and the Crypt

The microbiota is a complex ecosystem populated by a diverse community of microorganisms, mainly dominated by bacteria, but also comprised of archaea, fungi, protozoa, and viruses, that inhabit a specific area of the body (skin, oral cavity). The most studied is the intestinal microbiota, particularly abundant in the colon, in which Bacteroidetes and Firmicutes represent the dominant phyla.

Microbial colonization starts from birth where host, genetic, and environmental factors will determine the composition of adult gut microbiota that will remain relatively stable throughout life. But infections, antibiotic treatments, and diet are all possible causes for a shift in the microbiota composition and distribution in the different region of the gut, causing dysbiosis. Increasing clinical evidence links dysbiosis with various immune, metabolic, and neurological disorders in both intestinal and extraintestinal sites (Donaldson, Lee, & Mazmanian, 2016).

The dialogue between the microbiota and intestinal cells (immune and epithelial) is mainly due to the expression of innate immune receptors (PRRs) that recognize conserved bacterial motifs defined as "microbe-associated molecular patterns" (MAMPs) (Kawai & Akira, 2011). Principal members of the PRRs are the Toll-like receptors (TLRs) and the nucleotide oligomerization domain receptors (NODs). TLRs are transmembrane proteins located at the cellular plasma membrane or at the endosomal membranes, while NODs are cytosolic proteins (Sellge & Kufer, 2015). Upon MAMP recognition, PRRs initiate a signaling cascade that usually trigger to an inflammatory response, through the activation of the transcriptional factor NF-κB (nuclear factor-kappa B) responsible of the production of proinflammatory cytokines such as IL-1β, IL-6, or IFN-γ. Indeed, in the gut, PRRs seem to be crucial for bacterial–host communications and for maintaining intestinal homeostasis (Abreu, 2010).

The recognition of commensal microflora by *TLRs* expressed by epithelial cells is required for intestinal homeostasis and this interaction is critical for the protection against injury and associated mortality (Rakoff-Nahoum, Paglino, Eslami-Varzaneh, Edberg, & Medzhitov, 2004). Comparing MyD88 KO mice or TLR2 and/or TLR4 KO mice to wt mice, they showed that the activation of TLRs in the epithelial cells by the commensal microbiota is critical for protection against as DSS-induced damage. The mucosal injury mediated by this agent could be improved by oral administration of purified MAMPs, lipopolysaccharide (LPS) or lipoteichoic acid, respectively, ligands for TLR4 and TLR2. This was the first evidence of the effect of the microbiota, particularly the released MAMPs, in preserving gut homeostasis and promoting tissue repair and host survival upon recognition of bacterial products through PRRs.

A study from the Backhed's group in 2012 showed that the gut microbiota modulated host gene expression differently along the length of the gut, with a large set of genes regulated in the small intestine and fewer in the colon (Larsson et al., 2012). Moreover, the TLR-signaling adaptor MyD88 was mainly essential for microbiota-induced colonic expression of the antimicrobial genes *Reg3β* and *Reg3γ*. This study emphasized the role of the microbiota in host gene expression and how this modulation is linked to the intrinsic function of the different regions of the gut.

In a more recent study, the same group compared host gene expression in the presence or in the absence of the microbiota (conventional vs germ-free mice) not only over the length of the gut but also within the same region comparing crypts vs apical fractions (Sommer, Nookaew, Sommer, Fogelstrand, & Bäckhed, 2015) during the time course of colonization. They showed that about 10% of the host's transcriptome was microbially regulated, mainly including genes annotated with functions in immunity, cell proliferation, and metabolism. As expected, in the crypts the genes linked to proliferation genes were more induced than in the tips, but also functions related to protein biosynthesis. Metabolic processes were instead modulated by the microbiota in the tips of both ileum and colon.

The capacity of the microbiota to modulate gene expression could be also linked to the role of the microbiota on regulating epigenetics in the intestinal epithelium, providing an imprinting of their presence on the host. Alengath et al. have shown that the epigenome-modifying enzyme histone deacetylase 3 (HDAC3) has a central role in maintaining intestinal homeostasis, especially following damage and inflammation (Alenghat et al., 2013) and that this regulation is mediated by signals provided by the intestinal microbiota. Intestinal epithelial cell proliferation, epithelial barrier function, and Paneth cell number were all altered in mice lacking HDAC3 in epithelial cells. These parameters nearly returned to normal when animals were raised in GF conditions, showing that the presence of commensal bacteria facilitates the expression of networks of genes that regulate epithelial function and tissue homeostasis. The mechanisms through which this epigenetic regulation is driven are not yet deciphered, but these data show an integrated dialogue between the microbiota and the host gene expression.

One of the described functions of the microbiota in the symbiotic interaction with the host is the role in the metabolism. For instance, germ-free rodents require a higher caloric intake to maintain their weight than conventional mice (Wostmann, Larkin, Moriarty, & Bruckner-Kardoss, 1983). Indeed, the gut microbiota produces a variety of nutrients including short-chain fatty acids (SCFAs), such as propionate and butyrate, and vitamins that regulate host cell physiology. Sansonetti's group have shown that lactate and acetate from *Lactobacillus casei* and *Bifidobacterium breve*, an intestinal symbiont, negatively regulate the cell cycle of intestinal epithelial cell lines (Matsuki et al., 2013), indicating how SCFAs have a role also in the modulation of the homeostasis of the intestinal epithelium.

Recently, the Stappenbeck's group have provided evidence of the intricate coevolution between host and microbiota showing the effect of butyrate in the gut (Kaiko et al., 2016). Butyrate is a dominant energy source for differentiated colonocytes. On the other hand, butyrate, acting as HDAC inhibitor, is a potent suppressor of proliferation through a Foxo3-dependent mechanism, therefore inducing cell-cycle arrest. To reconcile how the high concentration of luminal butyrate (i.e., 5 mM in mouse and ~70 mM in humans (Kaiko et al., 2016)) did not perturb intestinal regeneration in homeostatic condition, they demonstrated that the crypt structure protects the intestinal stem/progenitor from the antiproliferative effect of butyrate. Colonocytes present at the top of the crypts actively used the butyrate as their energy source, protecting at the same time ISCs from the antiproliferative activity of this SCFA. Damages to the surface of the epithelium that alter the architecture of the crypts led to exposure of ISCs to butyrate, therefore inhibiting the proliferative activity and delaying wound repair.

2.3.2 Microbiota and ISCs

Until now, little is known regarding a direct interaction between microbial products and ISCs. This is mainly due to the current idea that the intestinal mucus layer provides a physical barrier to the luminal contents to the epithelium, particularly to the cells in the crypt. This physical limit is usually destroyed by pathogen that has specific mechanisms in order to pass the barrier, or in case of disease where alterations of the homeostasis create a loss of balance and bacteria could gain access to the bottom of the crypts. Recently, Pédron and collaborators have identified in cecal and colonic crypts of mice, a restricted set of bacteria that reside in this region at homeostasis (Pédron et al., 2012). Using a combination of laser capture microdissection of the crypt luminal space on intestinal section amplification of the variable sequences of genomic DNA encoding 16S rRNA, next-generation sequencing (454 pyrosequencing), and bioinformatic analysis, it was demonstrated that these bacterial clusters were dominated by operational taxonomic units corresponding to unexpected commensal microorganisms such as *Acinetobacter* and *Delftia/Comamonas*, in other words strictly aerobic, nonfermentative genera, respectively, belonging to the gamma- and beta-proteobacterial families. We may hypothesize that this "crypt-specific core microbiota" has been selected to survive in the crypt environment due to their role as a gatekeeper, preventing the proliferation of more aggressive symbiotic microorganism (i.e., pathobionts) (Chow & Mazmanian, 2010) and pathogen, and/or by providing "niche signals" to the ISCs.

The presence of bacteria inside the intestinal crypts and the possibility of diffusion of bacterial products from the lumen into the crypts open the question of crypt cells, and particularly ISCs could directly sensing signals from the microbiota. Neal et al. showed that TLR4 is mainly expressed by cells at the crypt bottom, and that 50% of the Lgr5 + cells express this receptor, indicating their capacity to sense bacterial LPS. Using in vivo model and in vitro organoids, it was showed that upon LPS stimulation, TLR4 activation reduced proliferation and increased apoptosis in ISC. Moreover, it was showed that the apoptotic signal was dependent on upregulation of PUMA through TRIF in an MyD88- and TNF-independent manner.

In the context of examining the role of the microbiota as niche for the ISCs, Nigro et al. have investigated the capacity of the ISCs to respond to MAMPs (Nigro, Rossi, Commere, Jay, & Sansonetti, 2014). The muramyl dipeptide (MDP), a component of peptidoglycan, had a protective effect on ISC, particularly upon oxidative stress. This effect was mediated by the expression of NOD2 receptor in ISCs. In this context, although also Paneth cells express NOD2, even if at lower level compare to ISCs, the protective effect was not mediated by these cells, indicating the direct interaction between microbial products and ISCs that are able to sense directly the MDP and activate the mechanism of protection. Therefore, it was showed that in a steady-state condition, the bacteria perhaps do not give any specific advantage to stem cells, as indicated by the fact that NOD2KO mice are viable and do not present any particular difference compared to wt mice. However, upon injury, the presence of the microbiota, particularly the released MDP, has a protective effect on ISCs, making them capable to escape from death.

These direct interactions between bacterial products and ISCs show how the microbiota could directly stimulate the fate of the ISCs, therefore being part of the ISC niche.

3. CONCLUSIONS

The intestinal epithelium is the most vigorously self-renewing tissue in adult mammals. This renewal process is dependent on a pool of ISCs present at the crypt bottom, identified by the Lgr5 marker. Lgr5 + ISCs give rise to proliferating progenitors that differentiate into an absorptive enterocyte lineage and in secretory lineages: goblet-, enteroendocrine-, Paneth-, and tuft cells. Stemness and survival of ISCs are supported by a niche, composed of Paneth cells, mesenchymal cells, and microbiota component and metabolites that provide a complex gradient of factors. Maintenance of ISCs is key to ensure epithelial homeostasis and efficient repair following tissue injuries by any type of stress, be it infectious, physical, or chemical. The signals provided by these different components of the ISC niche are essential to guide the ISC in the decision of survive, dye, proliferate, or differentiate.

Experimental conditions allowed to identify and isolate the single components of the ISC niche. However, it is likely that in vivo, the combination of these different signals is essential to ensure the maintenance of a healthy niche.

Moreover, different situations, i.e., homeostasis vs stress, make the different elements participating at different levels.

REFERENCES

Abreu, M. T. (2010). Toll-like receptor signalling in the intestinal epithelium: How bacterial recognition shapes intestinal function. *Nature Reviews Immunology*, *10*(2), 131–144.

Alenghat, T., Osborne, L. C., Saenz, S. A., Kobuley, D., Ziegler, C. G. K., Mullican, S. E., et al. (2013). Histone deacetylase 3 coordinates commensal-bacteria-dependent intestinal homeostasis. *Nature*, *504*(7478), 153–157.

Altmann, G. G. (1983). Morphological observations on mucus-secreting nongoblet cells in the deep crypts of the rat ascending colon. *The American Journal of Anatomy*, *167*(1), 95–117.

Aoki, R., Shoshkes-Carmel, M., Gao, N., Shin, S., May, C. L., Golson, M. L., et al. (2016). Foxl1-expressing mesenchymal cells constitute the intestinal stem cell niche. *Cellular and Molecular Gastroenterology and Hepatology*, *2*(2), 175–188.

Barker, N., van Es, J. H., Kuipers, J., Kujala, P., van den Born, M., Cozijnsen, M., et al. (2007). Identification of stem cells in small intestine and colon by marker gene Lgr5. *Nature*, *449*(7165), 1003–1007.

Bastide, P., Darido, C., Pannequin, J., Kist, R., Robine, S., Marty-Double, C., et al. (2007). Sox9 regulates cell proliferation and is required for Paneth cell differentiation in the intestinal epithelium. *The Journal of Cell Biology*, *178*(4), 635–648.

Beyaz, S., Mana, M. D., Roper, J., Kedrin, D., Saadatpour, A., Hong, S.-J., et al. (2016). High-fat diet enhances stemness and tumorigenicity of intestinal progenitors. *Nature*, *531*(7592), 53–58.

Birchenough, G. M. H., Johansson, M. E. V., Gustafsson, J. K., Bergström, J. H., & Hansson, G. C. (2015). New developments in goblet cell mucus secretion and function. *Mucosal Immunology*, *8*(4), 712–719.

Brown, S. L., Riehl, T. E., Walker, M. R., Geske, M. J., Doherty, J. M., Stenson, W. F., et al. (2007). Myd88-dependent positioning of Ptgs2-expressing stromal cells maintains colonic epithelial proliferation during injury. *The Journal of Clinical Investigation*, *117*(1), 258–269.

Buczacki, S. J. A., Zecchini, H. I., Nicholson, A. M., Russell, R., Vermeulen, L., Kemp, R., et al. (2013). Intestinal label-retaining cells are secretory precursors expressing Lgr5. *Nature*, *495*(7439), 65–69.

Caballero, S., & Pamer, E. G. (2015). Microbiota-mediated inflammation and antimicrobial defense in the intestine. *Annual Review of Immunology*, *33*, 227–256.

Chang, J. E., & Turley, S. J. (2015). Stromal infrastructure of the lymph node and coordination of immunity. *Trends in Immunology*, *36*(1), 30–39.

Cheng, H., & Leblond, C. P. (1974a). Origin, differentiation and renewal of the four main epithelial cell types in the mouse small intestine. I. Columnar cell. *The American Journal of Anatomy*, *141*(4), 461–479.

Cheng, H., & Leblond, C. P. (1974b). Origin, differentiation and renewal of the four main epithelial cell types in the mouse small intestine. V. Unitarian theory of the origin of the four epithelial cell types. *The American Journal of Anatomy*, *141*(4), 537–561.

Chow, J., & Mazmanian, S. K. (2010). A pathobiont of the microbiota balances host colonization and intestinal inflammation. *Cell Host & Microbe*, 7(4), 265–276.

Clevers, H. (2013). The intestinal crypt, a prototype stem cell compartment. *Cell*, *154*(2), 274–284.

Clevers, H. C., & Bevins, C. L. (2013). Paneth cells: Maestros of the small intestinal crypts. *Annual Review of Physiology*, *75*, 289–311.

de Lau, W., Barker, N., Low, T. Y., Koo, B.-K., Li, V. S. W., Teunissen, H., et al. (2012). Lgr5 homologues associate with Wnt receptors and mediate R-spondin signalling. *Nature*, *476*(7360), 293–297.

Donaldson, G. P., Lee, S. M., & Mazmanian, S. K. (2016). Gut biogeography of the bacterial microbiota. *Nature Reviews Microbiology*, *14*(1), 20–32.

Durand, A., Donahue, B., Peignon, G., Letourneur, F., Cagnard, N., Slomianny, C., et al. (2012). Functional intestinal stem cells after Paneth cell ablation induced by the loss of transcription factor Math1 (Atoh1). *Proceedings of the National Academy of Sciences of the United States of America*, *109*(23), 8965–8970.

Farin, H. F., van Es, J. H., & Clevers, H. (2012). Redundant sources of Wnt regulate intestinal stem cells and promote formation of Paneth cells. *Gastroenterology*, *143*(6), 1518–1529.e7.

Ferraro, F., Celso, C. L., & Scadden, D. (2010). Adult stem cells and their niches. *Advances in Experimental Medicine and Biology*, *695*, 155–168.

Garabedian, E. M., Roberts, L. J., McNevin, M. S., & Gordon, J. I. (1997). Examining the role of Paneth cells in the small intestine by lineage ablation in transgenic mice. *The Journal of Biological Chemistry*, *272*(38), 23729–23740.

Gerbe, F., & Jay, P. (2016). Intestinal tuft cells: Epithelial sentinels linking luminal cues to the immune system. *Mucosal Immunology*, *9*(6), 1353–1359.

Gribble, F. M., & Reimann, F. (2017). Signalling in the gut endocrine axis. *Physiology & Behavior*, *176*, 183–188.

Haramis, A.-P. G., Begthel, H., van den Born, M., van Es, J., Jonkheer, S., Offerhaus, G. J. A., et al. (2004). De novo crypt formation and juvenile polyposis on BMP inhibition in mouse intestine. *Science*, *303*(5664), 1684–1686.

He, X. C., Zhang, J., Wei-Gang, T., Tawfik, O., & Ross, J. (2004). BMP signaling inhibits intestinal stem cell self-renewal through suppression of Wnt-[beta]-catenin signaling. *Nature Genetics*, *36*(10), 1117–1121.

Horiguchi, H., Endo, M., Kawane, K., Kadomatsu, T., Terada, K., Morinaga, J., et al. (2017). ANGPTL2 expression in the intestinal stem cell niche controls epithelial regeneration and homeostasis. *The EMBO Journal*, *36*(4), 409–424.

Hsu, D. R., Economides, A. N., Wang, X., Eimon, P. M., & Harland, R. M. (1998). The Xenopus dorsalizing factor gremlin identifies a novel family of secreted proteins that antagonize BMP activities. *Molecular Cell*, *1*(5), 673–683.

Igarashi, M., & Guarente, L. (2016). mTORC1 and SIRT1 cooperate to foster expansion of gut adult stem cells during calorie restriction. *Cell*, *166*(2), 436–450.

Iwasaki, A., & Medzhitov, R. (2015). Control of adaptive immunity by the innate immune system. *Nature Immunology*, *16*(4), 343–353.

Kabiri, Z., Greicius, G., Madan, B., Biechele, S., Zhong, Z., Zaribafzadeh, H., et al. (2014). Stroma provides an intestinal stem cell niche in the absence of epithelial Wnts. *Development*, *141*(11), 2206–2215.

Kaiko, G. E., Ryu, S. H., Koues, O. I., Collins, P. L., Solnica-Krezel, L., Pearce, E. J., et al. (2016). The colonic crypt protects stem cells from microbiota-derived metabolites. *Cell*, *165*(7), 1708–1720.

Kawai, T., & Akira, S. (2011). Toll-like receptors and their crosstalk with other innate receptors in infection and immunity. *Immunity*, *34*(5), 637–650.

Kim, K.-A., Kakitani, M., Zhao, J., Oshima, T., Tang, T., Binnerts, M., et al. (2005). Mitogenic influence of human R-spondin1 on the intestinal epithelium. *Science*, *309*(5738), 1256–1259.

Korinek, V., Barker, N., Moerer, P., van Donselaar, E., Huls, G., Peters, P. J., et al. (1998). Depletion of epithelial stem-cell compartments in the small intestine of mice lacking Tcf-4. *Nature Genetics*, *19*(4), 379–383.

Kosinski, C., Li, V. S. W., Chan, A. S. Y., Zhang, J., Ho, C., Tsui, W. Y., et al. (2007). Gene expression patterns of human colon tops and basal crypts and BMP antagonists as intestinal stem cell niche factors. *Proceedings of the National Academy of Sciences of the United States of America*, *104*(39), 15418–15423.

Kuhnert, F., Davis, C. R., Wang, H.-T., Chu, P., Lee, M., Yuan, J., et al. (2004). Essential requirement for Wnt signaling in proliferation of adult small intestine and colon revealed by adenoviral expression of Dickkopf-1. *Proceedings of the National Academy of Sciences of the United States of America*, *101*(1), 266–271.

Larsson, E., Tremaroli, V., Lee, Y. S., Koren, O., Nookaew, I., Fricker, A., et al. (2012). Analysis of gut microbial regulation of host gene expression along the length of the gut and regulation of gut microbial ecology through MyD88. *Gut*, *61*(8), 1124–1131.

Lei, N. Y., Jabaji, Z., Wang, J., Joshi, V. S., Brinkley, G. J., Khalil, H., et al. (2014). Intestinal subepithelial myofibroblasts support the growth of intestinal epithelial stem cells. *PLoS One*, *9*(1), e84651.

Lindemans, C. A., Calafiore, M., Mertelsmann, A. M., O'Connor, M. H., Dudakov, J. A., Jenq, R. R., et al. (2015). Interleukin-22 promotes intestinal-stem-cell-mediated epithelial regeneration. *Nature*, *528*(7583), 560–564.

Madison, B. B., Braunstein, K., Kuizon, E., Portman, K., Qiao, X. T., & Gumucio, D. L. (2005). Epithelial hedgehog signals pattern the intestinal crypt-villus axis. *Development*, *132*(2), 279–289.

Mahapatro, M., Foersch, S., Hefele, M., He, G.-W., Giner-Ventura, E., Mchedlidze, T., et al. (2016). Programming of intestinal epithelial differentiation by IL-33 derived from pericryptal fibroblasts in response to systemic infection. *Cell Reports*, *15*(8), 1743–1756.

Manieri, N. A., Drylewicz, M. R., Miyoshi, H., & Stappenbeck, T. S. (2012). Igf2bp1 is required for full induction of Ptgs2 mRNA in colonic mesenchymal stem cells in mice. *Gastroenterology*, *143*(1), 110–121.e10.

Matsuki, T., Pédron, T., Regnault, B., Mulet, C., Hara, T., & Sansonetti, P. J. (2013). Epithelial cell proliferation arrest induced by lactate and acetate from Lactobacillus casei and Bifidobacterium breve. *PLoS One*, *8*(4), e63053.

Miller, H., Zhang, J., Kuolee, R., Patel, G. B., & Chen, W. (2007). Intestinal M cells: The fallible sentinels? *World Journal of Gastroenterology*, *13*(10), 1477–1486.

Molloy, M. J., Bouladoux, N., & Belkaid, Y. (2012). Intestinal microbiota: Shaping local and systemic immune responses. *Seminars in Immunology*, *24*(1), 58–66.

Mori Akiyama, Y., van den Born, M., van Es, J. H., Hamilton, S. R., Adams, H. P., Zhang, J., et al. (2007). SOX9 is required for the differentiation of paneth cells in the intestinal epithelium. *Gastroenterology, 133*(2), 539–546.

Muñoz, J., Stange, D. E., Schepers, A. G., van de Wetering, M., Koo, B.-K., Itzkovitz, S., et al. (2012). The Lgr5 intestinal stem cell signature: Robust expression of proposed quiescent "+4" cell markers. *The EMBO Journal, 31*(14), 3079–3091.

Nigro, G., Rossi, R., Commere, P.-H., Jay, P., & Sansonetti, P. J. (2014). The cytosolic bacterial peptidoglycan sensor Nod2 affords stem cell protection and links microbes to gut epithelial regeneration. *Cell Host & Microbe, 15*(6), 792–798.

Ootani, A., Li, X., Sangiorgi, E., Ho, Q. T., Ueno, H., Toda, S., et al. (2009). Sustained in vitro intestinal epithelial culture within a Wnt-dependent stem cell niche. *Nature Medicine, 15*(6), 701–706.

Pédron, T., Mulet, C., Dauga, C., Frangeul, L., Chervaux, C., Grompone, G., et al. (2012). A crypt-specific core microbiota resides in the mouse colon. *mBio, 3*(3), e00116–12.

Peduto, L., Dulauroy, S., Lochner, M., Späth, G. F., Morales, M. A., Cumano, A., et al. (2009). Inflammation recapitulates the ontogeny of lymphoid stromal cells. *The Journal of Immunology, 182*(9), 5789–5799.

Pinto, D., Gregorieff, A., Begthel, H., & Clevers, H. (2003). Canonical Wnt signals are essential for homeostasis of the intestinal epithelium. *Genes & Development, 17*(14), 1709–1713.

Potten, C. S., Hume, W. J., Reid, P., & Cairns, J. (1978). The segregation of DNA in epithelial stem cells. *Cell, 15*(3), 899–906.

Potten, C. S., Owen, G., & Booth, D. (2002). Intestinal stem cells protect their genome by selective segregation of template DNA strands. *Journal of Cell Science, 115*(Pt. 11), 2381–2388.

Powell, D. W., Pinchuk, I. V., Saada, J. I., Chen, X., & Mifflin, R. C. (2011). Mesenchymal cells of the intestinal lamina propria. *Annual Review of Physiology, 73*, 213–237.

Pull, S. L., Doherty, J. M., Mills, J. C., Gordon, J. I., & Stappenbeck, T. S. (2005). Activated macrophages are an adaptive element of the colonic epithelial progenitor niche necessary for regenerative responses to injury. *Proceedings of the National Academy of Sciences of the United States of America, 102*(1), 99–104.

Rakoff-Nahoum, S., Paglino, J., Eslami-Varzaneh, F., Edberg, S., & Medzhitov, R. (2004). Recognition of commensal microflora by toll-like receptors is required for intestinal homeostasis. *Cell, 118*(2), 229–241.

Rodríguez-Colman, M. J., Schewe, M., Meerlo, M., Stigter, E., Gerrits, J., Pras-Raves, M., et al. (2017). Interplay between metabolic identities in the intestinal crypt supports stem cell function. *Nature, 543*(7645), 424–427.

Rothenberg, M. E., Nusse, Y., Kalisky, T., Lee, J. J., Dalerba, P., Scheeren, F., et al. (2012). Identification of a cKit(+) colonic crypt base secretory cell that supports Lgr5(+) stem cells in mice. *Gastroenterology, 142*(5), 1195–1205.e6.

Sangiorgi, E., & Capecchi, M. R. (2008). Bmi1 is expressed in vivo in intestinal stem cells. *Nature Genetics, 40*(7), 915–920.

San Roman, A. K., Jayewickreme, C. D., Murtaugh, L. C., & Shivdasani, R. A. (2014). Wnt secretion from epithelial cells and subepithelial myofibroblasts is not required in the mouse intestinal stem cell niche in vivo. *Stem Cell Reports, 2*(2), 127–134.

Sasaki, N., Sachs, N., Wiebrands, K., Ellenbroek, S. I. J., Fumagalli, A., Lyubimova, A., et al. (2016). Reg4 + deep crypt secretory cells function as epithelial niche for Lgr5 + stem cells in colon. *Proceedings of the National Academy of Sciences of the United States of America, 113*(37), E5399–407.

Sato, T., van Es, J. H., Snippert, H. J., Stange, D. E., Vries, R. G., van den Born, M., et al. (2011). Paneth cells constitute the niche for Lgr5 stem cells in intestinal crypts. *Nature, 469*(7330), 415–418.

Sato, T., Vries, R. G., Snippert, H. J., van de Wetering, M., Barker, N., Stange, D. E., et al. (2009). Single Lgr5 stem cells build crypt-villus structures in vitro without a mesenchymal niche. *Nature, 459*(7244), 262–265.
Schewe, M., Franken, P. F., Sacchetti, A., Schmitt, M., Joosten, R., Böttcher, R., et al. (2016). Secreted phospholipases A2 are intestinal stem cell niche factors with distinct roles in homeostasis, inflammation, and cancer. *Cell Stem Cell, 19*(1), 38–51.
Schofield, R. (1978). The relationship between the spleen colony-forming cell and the haemopoietic stem cell. *Blood Cells, 4*(1–2), 7–25.
Seiler, K. M., Schenhals, E. L., von Furstenberg, R. J., Allena, B. K., Smith, B. J., Scaria, D., et al. (2015). Tissue underlying the intestinal epithelium elicits proliferation of intestinal stem cells following cytotoxic damage. *Cell and Tissue Research, 361*(2), 427–438.
Sellge, G., & Kufer, T. A. (2015). PRR-signaling pathways: Learning from microbial tactics. *Seminars in Immunology, 27*(2), 75–84.
Shroyer, N. F., Wallis, D., Venken, K. J. T., Bellen, H. J., & Zoghbi, H. Y. (2005). Gfi1 functions downstream of Math1 to control intestinal secretory cell subtype allocation and differentiation. *Genes & Development, 19*(20), 2412–2417.
Smith, N. R., Gallagher, A. C., & Wong, M. H. (2016). Defining a stem cell hierarchy in the intestine: Markers, caveats and controversies. *The Journal of Physiology, 594*(17), 4781–4790.
Sommer, F., Nookaew, I., Sommer, N., Fogelstrand, P., & Bäckhed, F. (2015). Site-specific programming of the host epithelial transcriptome by the gut microbiota. *Genome Biology, 16*, 62.
Stzepourginski, I., Nigro, G., Jacob, J.-M., Dulauroy, S., Sansonetti, P. J., Eberl, G., et al. (2017). CD34+ mesenchymal cells are a major component of the intestinal stem cells niche at homeostasis and after injury. *Proceedings of the National Academy of Sciences of the United States of America, 114*(4), E506–E513.
Tetteh, P. W., Basak, O., Farin, H. F., Wiebrands, K., Kretzschmar, K., Begthel, H., et al. (2016). Replacement of lost Lgr5-positive stem cells through plasticity of their enterocyte-lineage daughters. *Cell Stem Cell, 18*(2), 203–213.
van Dop, W. A., Uhmann, A., Wijgerde, M., Sleddens Linkels, E., Heijmans, J., Offerhaus, G. J., et al. (2009). Depletion of the colonic epithelial precursor cell compartment upon conditional activation of the hedgehog pathway. *Gastroenterology, 136*(7), 2195–2203.e7.
van Es, J. H., Sato, T., van de Wetering, M., Lyubimova, A., Nee, A. N. Y., Gregorieff, A., et al. (2012). Dll1+ secretory progenitor cells revert to stem cells upon crypt damage. *Nature Cell Biology, 14*(10), 1099–1104.
van Es, J. H., van Gijn, M. E., Riccio, O., van den Born, M., Vooijs, M., Begthel, H., et al. (2005). Notch/[gamma]-secretase inhibition turns proliferative cells in intestinal crypts and adenomas into goblet cells. *Nature, 435*(7044), 959–963.
Walker, M. R., Brown, S. L., Riehl, T. E., Stenson, W. F., & Stappenbeck, T. S. (2010). Growth factor regulation of prostaglandin-endoperoxide synthase 2 (Ptgs2) expression in colonic mesenchymal stem cells. *Journal of Biological Chemistry, 285*(7), 5026–5039.
Wilm, B., Ipenberg, A., Hastie, N. D., Burch, J. B. E., & Bader, D. M. (2005). The serosal mesothelium is a major source of smooth muscle cells of the gut vasculature. *Development, 132*(23), 5317–5328.
Worthley, D. L., Churchill, M., Compton, J. T., Tailor, Y., Rao, M., Si, Y., et al. (2015). Gremlin 1 identifies a skeletal stem cell with bone, cartilage, and reticular stromal potential. *Cell, 160*(1–2), 269–284.
Wostmann, B. S., Larkin, C., Moriarty, A., & Bruckner-Kardoss, E. (1983). Dietary intake, energy metabolism, and excretory losses of adult male germfree Wistar rats. *Laboratory Animal Science, 33*(1), 46–50.
Yilmaz, Ö. H., Katajisto, P., Lamming, D. W., Gültekin, Y., Bauer-Rowe, K. E., Sengupta, S., et al. (2012). mTORC1 in the Paneth cell niche couples intestinal stem-cell function to calorie intake. *Nature, 486*(7404), 490–495.

CHAPTER FOUR

Extracellular Matrix Remodeling in Intestinal Homeostasis and Disease

Sang-Eun Lee[*,2], **Isobel Massie**[*,2], **Laween Meran**[*,†], **Vivian S.W. Li**[*,1]

[*]Stem Cell and Cancer Biology, The Francis Crick Institute, London, United Kingdom
[†]Stem Cells and Regenerative Medicine Section, University College London Institute of Child Health, London, United Kingdom
[1]Corresponding author: e-mail address: vivian.li@crick.ac.uk

Contents

1. Mammalian Intestine and the Surrounding Niche 100
   1.1 Intestinal Stem Cells 100
   1.2 The Stem Cell Niche 101
   1.3 Mesenchymal Cells in the ISC Niche 103
   1.4 Overview of ECM in the Intestine 104
   1.5 Reconstitution of ISC Niche Ex Vivo 105
2. ECM Dynamics in Intestinal Homeostasis and Regeneration 105
   2.1 ECM Remodeling in Homeostasis 105
   2.2 Role of ECM Niche in ISC Fate Control 107
   2.3 ECM Remodeling Upon Injury and Regeneration 108
3. ECM Dynamics in IBD 111
   3.1 Inflammatory Bowel Disease 111
   3.2 ECM Composition in IBD 113
   3.3 ECM Remodeling in IBD 115
4. ECM Dynamics in CRC 115
   4.1 CRC Microenvironment 115
   4.2 ECM Remodeling for Invasion and Metastasis 120
   4.3 CRC Therapeutic Strategies Targeting Tumor Microenvironment 121
5. Tissue Engineering 122
   5.1 Intestinal Tissue Engineering 122
   5.2 CRC Modeling 124
6. Conclusion 127
Acknowledgments 128
References 128

[2] These authors contributed equally to the work.

Advances in Stem Cells and their Niches, Volume 2
ISSN 2468-5097
https://doi.org/10.1016/bs.asn.2018.01.001

1. MAMMALIAN INTESTINE AND THE SURROUNDING NICHE

The mammalian intestine consists of the small intestine (duodenum, jejunum, and ileum) and the large intestine (colon and rectum). Food digestion and nutrient absorption primarily takes place in the small intestine, whereas water resorption occurs mostly in the colon. The lumen of the intestine is covered by a layer of intestinal epithelial cells that are organized into crypt–villus units. Crypt invaginations from the mucosal surface into the deeper mesenchymal layer, also termed crypts of Lieberkühn, are present throughout small intestine and colon, while protruding villi are found only in small intestine. The intestinal epithelium is one of the most vigorously regenerating tissues. The sheet of cells covering the entire intestinal lumen, which is approximately the size of a tennis court, is continuously renewed every 4–5 days. Such a high turnover rate is supported by the stem cells located at the bottom of the proliferative crypts. Intestinal stem cells (ISCs) divide and give rise to the trans-amplifying (TA) progenitor cells, which migrate upward for differentiation toward the villi of the small intestine or upper crypts of the colon. The differentiated cells include enterocytes, goblet cells, enteroendocrine cells, tuft cells, and Paneth cells. Paneth cells are the only cell type that migrate downward toward the crypt bottom. ISC homeostasis is tightly regulated by the surrounding stem cell niche, which includes both mesenchymal cells and the extracellular matrix (ECM). In this chapter, we discuss how stem cell niche contributes to intestinal homeostasis and disease with the focus on ECM remodeling, and provide an overview of the current advances in the field of tissue engineering.

1.1 Intestinal Stem Cells

ISCs are located at the base of intestinal crypts to fuel the rapid renewing epithelium. Thus far, two models of ISC location have dominated the field: the "+4 position" and the crypt base columnar (CBC) cells. The classic +4 stem cells (counting cell nuclei from the crypt bottom) were first proposed in 1974 based on a DNA label retaining study (Potten, Kovacs, & Hamilton, 1974). These cells are generally considered as quiescent with the potential to activate upon injury. The most notable marker for the +4 cells is Bmi1 that can give rise to all cell lineages (Sangiorgi & Capecchi, 2008). The second ISC model was the slender CBC cells intermingled with Paneth cells at the crypt bottom (Cheng & Leblond, 1974a, 1974b). The leucine-rich repeat-containing

G protein-coupled receptor 5 (Lgr5) was the first CBC-specific marker identified in 2007 based on lineage tracing analysis (Barker et al., 2007). Unlike the +4 stem cells, the Lgr5-expressing CBC cells are highly proliferative with a cell cycle of 12–24 h. Both +4 and CBC stem cells were shown to be long-lived and multipotent with the ability to regenerate all different cell types in the epithelium. Interestingly, Bmi1-expressing cells were shown with the ability to replenish the Lgr5-expressing cells when these cycling CBC cells were selective killed (Tian et al., 2012). The data suggest that Bmi1-cells may be the reserve stem cell population for Lgr5-cells upon damage.

1.2 The Stem Cell Niche

ISC homeostasis is strictly regulated by multiple signaling pathways to control cell proliferation and differentiation. The major pathways that regulate these processes include Wnt and its agonist R-spondin (Rspo), Notch, and bone morphogenic protein (BMP). Wnt signaling is activated at the crypt bottom and is crucial to maintain ISC self-renewal and proliferation (Schuijers & Clevers, 2012). A more recent study demonstrates that Wnt confers a priming function to the ISCs by inducing Rspo receptor expression, thereby enabling Rspo-mediated ISC self-renewal (Yan et al., 2017). BMP signaling, on the other hand, is activated at the villus to inhibit ISC expansion and promote cell differentiation (He et al., 2004; Qi et al., 2016). Notch signaling is known to play a critical role in lineage commitment of the TA progenitor cells. Notch activation is required for absorptive lineage differentiation, while suppression of Notch will drive secretory lineage differentiation (van Es et al., 2005). This complicated signaling network is established by morphogen gradients along the crypt–villus axis in the surrounding niche (see Section 1.3).

The ISC niche consists of both cellular and physical components. The cellular niche contains various mesenchymal cells that secrete essential growth factors to maintain ISC homeostasis (see later). In addition to the stromal microenvironment, Paneth cells within close proximity to ISCs at the crypt base were recently recognized as ISC niche. Paneth cells secrete not only antimicrobial peptides to protect ISC from the hostile environment, but also crucial factors such as Wnt proteins, epidermal growth factor (EGF), and Notch ligands required for stem cell maintenance (Sato, van Es, et al., 2011). Reg4-expressing secretory cells are thought to be the equivalent niche cells in colon (Sasaki et al., 2016). The physical niche refers to the ECM components in the lamina propria, which support ISC homeostasis via biochemical and mechanical signals (Fig. 1A).

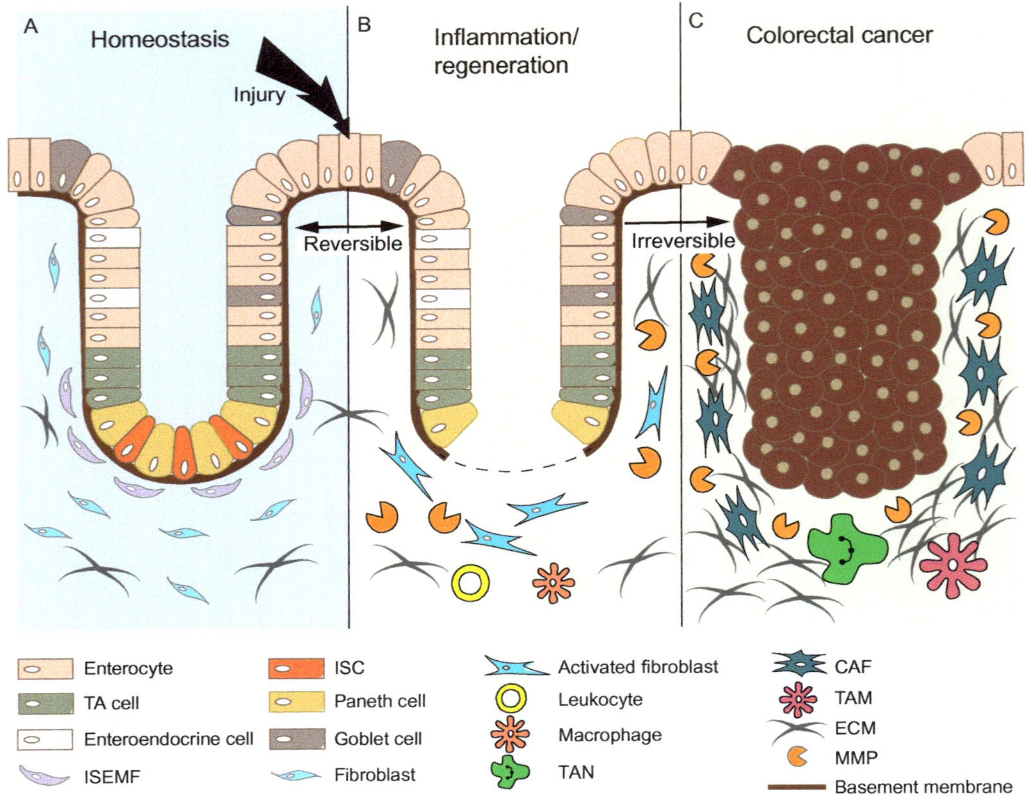

Fig. 1 See legend on opposite page.

1.3 Mesenchymal Cells in the ISC Niche

The mesenchymal cells of the intestine include fibroblasts, intestinal subepithelial myofibroblasts (ISEMFs), smooth muscle cells, endothelial cells, neural cells, and immune cells (reviewed in Sailaja, He, & Li, 2016). ISEMFs can be distinguished from fibroblasts and smooth muscle cells by their coexpression of α-smooth muscle actin (αSMA) and vimentin, and absence of muscle-specific marker desmin (Jabaji et al., 2014). Notably, the pericryptal ISEMFs are located within the lamina propria in close proximity to the ISCs and are thought to provide signaling cues in the stromal niche for ISC homeostasis by establishing morphogen gradients.

For instance, several Wnt ligands are secreted by stromal cells at the crypt to maintain the Wnt gradient radiating from the crypt bottom (Farin, Van Es, & Clevers, 2012; Gregorieff et al., 2005). Conversely, BMP4 is expressed by stromal cells within the lamina propria at the villus tip to promote epithelial cell differentiation (He et al., 2004). The BMP antagonists Gremlins-1 and -2 and Chordin, on the other hand, are secreted by the ISEMFs and smooth muscle cells to antagonize BMP signal at the crypt bottom for ISC maintenance (Kosinski et al., 2007). A recent publication has identified yet another population of $CD34^+/\alpha SMA^-$ mesenchymal cells (i.e., a population distinct from αSMA^+ ISEMFs) that secrete Wnt2b, Gremlin-1, and R-spondin 1 at the ISC niche (Stzepourginski et al., 2017). Another key signaling pathway that involves the stromal cells is the Hedgehog signaling pathway. The two ligands Sonic Hedgehog and Indian Hedgehog are both expressed at the intervillus region of the

Fig. 1 Alterations in ECM and stromal cell composition in homeostasis, active inflammation/regeneration, and colorectal cancer. (A) During homeostasis, the basement membrane is intact and continuously renewed. Pericryptal ISEMFs are located in the stroma in close proximity to ISCs. Stromal fibroblasts maintain the ECM. (B) During inflammation and regeneration, the basement membrane is interrupted leading to an infiltration of immune cells, such as leukocytes and macrophages. This triggers activation of stromal fibroblasts and release of MMPs and other ECM peptidases that act to remodel the ECM. This process is reversible and once complete, normal homeostasis resumes. (C) In colorectal cancer, the basement membrane is breached by hyperproliferative epithelial cells. This is associated with increased abundance of cancer-associated fibroblasts and tumor-associated neutrophils and macrophages. Expression of MMPs and other ECM peptidases is also upregulated resulting in stiffening of the matrix. Collagen fibers are also aligned as a consequence of ECM remodeling. This process is irreversible. *CAF*, cancer-associated fibroblast; *ECM*, extracellular matrix; *ISC*, intestinal stem cell; *ISEMF*, intestinal subepithelial myofibroblast; *MMP*, matrix metalloproteinase; *TA cell*, transient amplifying cell; *TAM*, tumor-associated macrophage; *TAN*, tumor-associated neutrophil.

epithelium, while the receptors Patched and the downstream effectors Gli1, 2, and 3 are expressed in the surrounding mesenchyme (Madison et al., 2005). Together, the data suggest the crucial role of the intestinal stromal cells in regulating the signaling gradient at the niche for ISC homeostasis. In addition to the morphogen gradient role, the intestinal stromal cells are also thought to contribute to the deposition of multiple key ECM proteins including collagen type IV, laminins-β1 and -γ1, and fibronectin (Mahida, Beltinger, et al., 1997; Okuno et al., 2002).

1.4 Overview of ECM in the Intestine

ECM comes in two major forms: first, connective tissue matrix that forms the bulk of the tissue mass and provides tissue structure; second, basement membrane that separates epithelial cells from the underlying stroma. ECM proteins include laminins, collagens, heparin sulfate proteoglycans (HSPGs), and fibronectin, all of which are ligands for the cell surface receptors such as integrins (that physically connect the cytoskeleton of the cell to the ECM), and are important throughout both development and homeostasis (Simo et al., 1991).

Type IV collagen, laminin, and HSPG are found homogenously throughout crypt base to villus tip in the lamina propria (Beaulieu, 1992), although a later investigation revealed that laminin subunits are nonuniformly distributed along the basement membrane between crypts and villi (Beaulieu & Vachon, 1994). Fibronectin is secreted by epithelial cells and has been found to be present in the basement membrane lining at the crypts in both rat and human intestine (Gagne et al., 2010; Quaroni, Isselbacher, & Ruoslahti, 1978). Tenascin, on the other hand, is strongly expressed at the crypt base and villus tip but only weakly in between. Integrin expression in epithelial cells varies among different subunits—most of them express differentially between crypt and villus, while the α2 subunit is confined to the crypts (Beaulieu, 1992). Importantly, changes in the expression of some integrins by epithelial cells have been suggested to correlate with the onset of differentiation, indicating the pivotal role of ECM in ISC homeostasis (Basora, Vachon, Herring-Gillam, Perreault, & Beaulieu, 1997).

ECM proteins have also been shown to have an impact on cell phenotype in vitro. Various studies have demonstrated that the proliferation, differentiation, and function of Caco-2 cells can be affected by the composition of the ECM on which they are cultured (Basson, Turowski, & Emenaker, 1996; Wang, Murthy, Barabino, & Carrier, 2010; Wolpert, Wong, & Bass, 1996). Additionally, the ECM can act as a reservoir for various soluble growth factors that bind to the ECM proteins (particularly to HSPGs) in their latent forms, which can be released upon matrix remodeling (reviewed in Hynes, 2009). This highlights the requirement for the delicate control of ECM remodeling in tissue homeostasis.

1.5 Reconstitution of ISC Niche Ex Vivo

Several intestinal epithelial culture methods have been described that aim to reconstitute the ISC niche in vitro. Ootani et al. generated intestinal spheres that contain both the epithelial and mesenchymal cells derived from small intestine or colon fragments in a 3-dimensional (3D) collagen I matrix with an air–liquid interface (Ootani et al., 2009). Proliferative epithelial cells derived from tissue fragments were capable of long-term culture in tandem with myofibroblasts lining their basal surface, and contained Lgr5 + cells responsive to Wnt agonists and antagonists as expected of ISCs. This long-term primary ISC culture methodology indicates that the ex vivo 3D microenvironment can successfully recapitulate in vivo organization and facilitate cell–cell and cell–ECM interactions of the ISC niche.

Another ex vivo intestinal organoid system describes a means of culturing ISCs within a biomimetic environment that permits formation of 3D organoids comprising both ISCs and their more differentiated progeny including Paneth cells. Of note, these organoids are epithelium-only cultures maintained in laminin-rich Matrigel droplets in the absence of mesenchymal cells. This is achieved by reconstituting the ISC niche signal using growth factors including EGF, the BMP antagonist Noggin, and Wnt agonist Rspo (Sato et al., 2009). Intestinal organoids can be derived from both healthy intestinal tissue and colorectal cancers (CRCs), and have been proposed as model systems for healthy and diseased states, respectively (Sato, Stange, et al., 2011). These organoids have been demonstrated to respond to ECM cues since modification of biologically inert poly(ethylene glycol) (PEG) hydrogels with ECM proteins such as fibronectin, collagen IV, laminin 111, and perlecan, improved organoid formation efficiency (Gjorevski et al., 2016). The latter study further highlights the crucial role of ECM remodeling in ISC maintenance and differentiation.

2. ECM DYNAMICS IN INTESTINAL HOMEOSTASIS AND REGENERATION

2.1 ECM Remodeling in Homeostasis

The ECM is constantly undergoing dynamic remodeling during homeostasis. There are a number of enzymes involved in this, all of which act in harmony to ensure that the ECM is remodeled appropriately. Matrix metalloproteinases (MMPs) are the major group of enzymes acting to remodel the ECM. These proteins belong to a large family of zinc-dependent endopeptidases that are found mostly as inactive proforms during normal homeostasis. Twenty-three human MMPs had been identified as of 2014, each of which acts upon a specific ECM substrate (which may also involve other MMPs) (reviewed in Bonnans, Chou, & Werb, 2014). Cleavage and degradation of all ECM proteins are possible when all MMPs are activated. This results in loosening of the physically dense ECM that enables cellular migration throughout matrices and release of the bound growth factors in the ECM reservoir (reviewed in Bonnans et al., 2014). Similarly, heparanase cleaves HSPG and is considered to be a remodeling enzyme despite the weak expression in some intestinal epithelial cells during homeostasis (Friedmann et al., 2000).

ECM composition is crucial for many cellular processes. The MMP expression is tightly regulated by enzymes such as tissue-inhibitor of metalloproteinases (TIMPs) and adamalysins during normal homeostasis, and can also be affected by external stimuli such as mechanical strain and growth factors (reviewed in Bonnans et al., 2014 and Wang & Yan, 2006). Dysregulation of MMP activity is associated with diseases such as inflammation and cancer. Insufficient deposition of ECM proteins can lead to loss of tissue mass, while excessive deposition can cause tissue stiffening and fibrosis, which will be discussed in Section 3.2. MMP expression and gelatinolytic activity appear to differ between small and large intestines, both in epithelial cells and in the lamina propria (Solli, Fadnes, Winberg, Uhlin-Hansen, & Hadler-Olsen, 2013). For instance, MMPs (notably -2 but also -1, -3, and -9) and TIMP-1 are secreted by pericryptal ISEMFs in the absence of exogenous factors, suggesting the essential role of the mesenchymal cells in ECM remodeling during homeostasis (Bamba et al., 2003; Bonnans et al., 2014; Okuno et al., 2002).

Lysyl oxidase (LOX) and lysyl oxidase-like (LOXL) enzymes are copper-dependent oxidative enzymes that modify lysyl residues on protein substrates and crosslink collagen and elastin (Lucero & Kagan, 2006), thus affecting tissue stiffness (which will be discussed in more detail in Section 4.1). LOX and LOXL enzymes are expressed differently in small intestine and colon (with LOX nearly absent from colon) during homeostasis, although the functional relevance for these differences remains unclear (Hayashi et al., 2004).

Integrins have been shown to be important to permit continuous ECM deposition during homeostasis to balance tissue remodeling by MMPs and other ECM peptidases. Expression of the integrin-linked kinase (ILK) has been shown to be enriched at the basal plasma membrane of the epithelial cells located at the base of intestinal crypts. ILK expression in Caco-2 cells in vitro declines as they differentiate and appears to be necessary for normal cell spreading, migration, and basement membrane deposition (Gagne et al., 2010). These findings support the notion that the lack of a physical link between cells and the surrounding ECM can result in anoikis-mediated cell death (at the villus tip for example). Integrins also appear to be important for crypt fission, which occurs throughout life although with decreasing rates of incidence. It has recently been proposed that Paneth cells in the niche display increased expression of the β4 integrin compared to ISCs and are more adherent to the underlying laminin, resulting in a "stiffer" region. The intestinal crypts expand over time when ISCs divide, leading to a localized accumulation of ISCs. The relatively softer ISC-rich region will then initiate bifurcation of the crypt in response to pressure from the surrounding tissue, which marks the onset of crypt fission (Langlands et al., 2016).

The physical aspects of the ECM such as stiffness and topography are equally important but are often overlooked. ISCs are known to respond to physical as well as biochemical cues from the ECM. For instance, the intestinal organoid formation efficiency, shape, and the subcellular localization of the effector protein yes-associated protein (YAP) were affected when ISCs were embedded within biologically inert PEG hydrogels of varying stiffness (Gjorevski et al., 2016). Furthermore, stress induced by artificial matrix deformation has also been reported to stimulate a hyperproliferative response in normal epithelial cells in a mouse model (Fernandez-Sanchez et al., 2015). Similarly, topography has been shown to affect epithelial (Caco-2) cell spreading and barrier functionality in vitro (Wang et al., 2010). Therefore, the ECM at the ISC niche must be tightly regulated during homeostasis by the precise control of ECM peptidases and their inhibitors for proper cellular proliferation and differentiation.

2.2 Role of ECM Niche in ISC Fate Control

Small intestine and colon differ in a number of ways, notably in that villi and lysozyme-secreting Paneth cells are absent from colonic crypts. However, it remains unclear whether the region specification is determined intrinsically by the ISC identity or is influenced by the surrounding niche. Several studies have attempted to address this question using organoid transplantation approach, while the results were somewhat inconclusive. As one might expect, colonic organoids transplanted onto damaged colon were able to repopulate colonic crypts and form a normally differentiated colonic epithelial layer (Yui et al., 2012). On the other hand, when organoids derived from adult small bowel were transplanted onto damaged colon, the engrafted epithelium retained the small intestinal stem cell identity with functional Paneth cells and did not express colonic epithelial cell markers such as carbonic anhydrase-2 (CA2) (Fukuda et al., 2014). In some locations, "finger-like projections" appeared at 4 weeks posttransplantation resembling the typical small intestinal architecture. These results suggest that the intrinsic program of the ISCs dominates cell fate control over the surrounding niche. In contrast, transplantation of organoids derived from fetal small intestine onto damaged colon was able to regenerate colonic epithelium with CA2 expression, indicating an increased degree of plasticity of ISCs during early development (Fordham et al., 2013). However, it is unclear if the reprogramming of the ISC fate by the surrounding niche is caused by stromal cells or ECM.

Another study utilizing a different approach by modifying the region-specific laminin may provide some insight into the role of ECM in ISC fate. Laminin subtypes are differentially distributed along the basement membrane of the small intestine and colon. In particular, laminin-α5 is enriched in the villus basement membrane. Depletion of laminin-α5 resulted in transformation from small intestinal to colonic architecture with villus coalescence and colonic-like goblet cells (Mahoney, Stappenbeck, & Miner, 2008). However, expression of small intestinal marker sucrase isomaltase was retained, indicating the incomplete transformation of the epithelial cell identity. Despite the partial reprogramming of the region-specific cell identity, this is a solid example demonstrating the ability of the ISC fate control by ECM remodeling in vivo. However, ECM comprises only one aspect of the ISC niche and it is highly likely that the stromal cells and their site-specific secretory profile may also contribute a significant role in epithelial differentiation. It would be important to further explore the combinatory role of both ECM and stromal cells in the niche for ISC fate control.

2.3 ECM Remodeling Upon Injury and Regeneration

Intestinal injury may occur due to infection, irradiation, ischemia, or other similar insults. Dysregulated intestinal regeneration may significantly decrease the absorptive and barrier function of the epithelium, leading to severe chronic diseases such as inflammatory bowel disease (IBD) or even intestinal failure. Various models of acute intestinal injury have been developed to investigate how the intestine is regenerated. These include:

- radiation, which triggers apoptosis in proliferating ISCs
- biopsy, a localized mechanical injury that results in loss of luminal epithelium
- dextran sodium sulfate (DSS) administration, a chemical injury that causes ulceration and inflammation of the epithelium along the intestine as a colitis model
- introduction of bacteria, used as a model of infection-induced injury

Loss of epithelial integrity via interruption of the epithelial monolayer and breach of basement membrane can expose the underlying stromal cells to luminal contents such as microbes. Biopsy-induced injury causes a rapid influx of platelets, neutrophils, and red blood cells in colon (Seno et al., 2009). This is followed by migration of differentiated epithelial cells from adjacent uninjured areas along wound channels to the injury site so that the wound bed is covered with postmitotic, Claudin-4+ "wound-associated epithelial (WAE)" cells within approximately 24 h. These WAE cells are then replaced by mitotic epithelial cells from neighboring uninjured crypts by 4 days postinjury (Miyoshi, Ajima, Luo, Yamaguchi, & Stappenbeck, 2012; Seno et al., 2009). Around 6 days postinjury, a population of poorly differentiated Wnt5a-expressing mesenchymal cells are localized to the underlying stroma of the wound bed to inhibit epithelial cell proliferation locally via TGF-β pathway, thus permitting invagination of the stroma and de novo crypt formation 8–14 days after insult (Miyoshi et al., 2012).

Numerous stromal cell types are involved in repair following injury to the intestinal epithelium. Alterations in the expression of various cytokines, MMPs, and growth factors are important to ensure regeneration of healthy tissue is achieved. In an elegant organ culture study, the intestine was denuded to allow visualization of the basement membrane. Migration of lymphocytes, macrophages, and eosinophils was observed over 48 h via discrete pore in the basement membrane using electron microscopy (Mahida, Galvin, et al., 1997). Infiltration of immune cells indeed occurs sooner in vivo after around 6 h postbiopsy injury (Seno et al., 2009). Macrophages appear to be particularly important for epithelial regeneration. Absence of macrophages failed to initiate the normal epithelial proliferative response to injury, while proliferation was still observed when lymphocytes and neutrophils were depleted (Pull, Doherty, Mills, Gordon, & Stappenbeck, 2005). Interestingly, macrophages appear to be recruited to laminin-deficient areas, suggesting that the interruption of the basement membrane may trigger this recruitment.

Increased concentrations of TNF-α, IFN-γ, and interleukins were observed following insult (Nava et al., 2010; Song et al., 2015), resulting in upregulated MMP expression in stromal cells such as ISEMFs (Bamba et al., 2003; Okuno et al., 2002; Pull et al., 2005). In a radiation-induced injury model, a rapid (within 1 day) increase of MMP-3 expression in infiltrating macrophages as well as MMP-9 from unspecified source was observed. This was followed by increased MMP-2 expression in ISEMFs and epithelial cells at 3 days postinjury (Strup-Perrot, Vozenin-Brotons, Vandamme, Benderitter, & Mathe, 2006). Similarly, in a DSS injury model, expression levels of MMP-2 and -9 were increased 2 days following cessation of DSS exposure (Breynaert et al., 2016). Expression of TIMP-1 and -2 was also found upregulated at the restitution phase, which may act to inhibit the degrading action of the MMPs during acute wound healing response (Strup-Perrot et al., 2006). Interestingly, DSS-injured TIMP-1 knockout animals recovered significantly faster than wildtype despite the initial increase in MMP activity and inflammation, suggesting that the balance between MMPs and their inhibitors is complex and critical for intestinal regeneration (Breynaert et al., 2016). Proper function of integrins is also required to aid epithelial migration during regeneration. In particular, α6- and α3β1-integrins were shown to be important for reepithelialization (Lotz et al., 1997). Interestingly, MMPs can also indirectly regulate integrin affinity by shedding HSPGs to aid this process (Chen et al., 2009).

Cytokine release can also differentially alter ECM deposition. For instance, IL-1β, TNF-α, and TGF-β can all trigger increased secretion of collagen types I and IV by ISEMFs, while only IL-1β induces fibronectin secretion (Okuno et al., 2002). Similarly, TGF-β1 has been shown to induce enhanced procollagen I and III deposition as well as other collagens via Smad3 signaling by intestinal fibroblasts (Luna, Masamunt, Llach, Delgado, & Sans, 2011; Verrecchia, Chu, & Mauviel, 2001).

As mentioned earlier, the ECM can act as a reservoir for growth factors and cytokines. These proteins may be released or activated upon MMP-mediated proteolysis of the ECM following injury (reviewed in Hynes, 2009). MMPs can also bind to ECM proteins, making this process self-perpetuating (Shi, Ramsay, Ermis, & Carson, 2012). Simultaneously, remodeling of the ECM enhances cell migration, which can result in increased localized concentrations of growth factors as a consequence of mesenchymal cell infiltration. Interestingly, laminin has been shown to inhibit expression of both insulin-like growth factor 1 (IGF1) and EGF receptors in cultured IEC-6 cells, suggesting that loss of contact between epithelial cells and their basement membrane proteins "primes" ISCs to respond to these factors (Wolpert, Wong, Wang, & Bass, 1996). Both EGF and IGF-1 have been shown to promote epithelial proliferation and ISC survival following radiation-induced injury, resulting in increased crypt density and depth (Potten et al., 1995; Van Landeghem et al., 2015; Wolpert et al., 1996). Similarly, fibroblast growth factors (FGFs) expression in the ECM niche surrounding intestinal crypts was induced after irradiation, while FGF2 expression was observed in immune cells preceding epithelial repair (Houchen, George, Sturmoski, & Cohn, 1999). Wnt signaling is often activated during intestinal regeneration, while ECM may contribute to one potential source of Wnt proteins during remodeling (Yue, 2014). Expression of Wnt antagonist Dkk1 is altered in dendritic cells, lymphocytes, and platelets (and to a lesser degree macrophages) following DSS injury in mice, leading to increased proliferation of ISCs via enhanced Wnt/β-catenin signaling and a more rapid recovery in an inflammation model (Koch et al., 2011; Nava et al., 2010; Song et al., 2015). The Wnt agonists, R-spondins, are also upregulated differentially in the stromal niche depending on the cause of injury. For instance, R-spondin 1 was increased following irradiation while R-spondin 2 was elevated after infection (Bhanja et al., 2009; Kang, Yousefi, & Gruenheid, 2016). A more recent study further showed that the expression of R-spondin 1 and the BMP antagonist Gremlin-1 was upregulated in a stromal population surrounding the crypts upon injury (Stzepourginski et al., 2017). Together, the data suggest the importance of growth factor modulation upon injury by both ECM remodeling and stromal cell infiltration.

In summary, ECM remodeling plays an indispensable role in intestinal regeneration. Disruption of the basement membrane triggers recruitment of immune cells to release various cytokines that alter the secretory profile of the stromal cells. Activation of ECM peptidases allows recruitment and homing of cells to the site of injury to repair the ECM by depositing new matrix proteins. Simultaneously, ECM-bound factors are released to promote epithelial cell migration and proliferation for rapid restoration of epithelial integrity (Fig. 1B).

3. ECM DYNAMICS IN IBD

3.1 Inflammatory Bowel Disease

Acute inflammatory episodes affecting the gut are relatively self-limiting, which evoke normal tissue regenerative responses for wound healing lasting for hours to days and return to homeostasis (discussed earlier). In contrast, chronic inflammatory disorders can lead to long-term irreversible modifications of the intestinal ECM (Rieder, Brenmoehl, Leeb, Scholmerich, & Rogler, 2007; Scharl & Rogler, 2014). IBD is a chronic relapsing and remitting inflammatory condition characterized by an excessive immune response and the associated tissue damage to the intestine, as well as a reduced diversity of commensal gut microbiota (Baumgart & Sandborn, 2012; Hanauer, 2004; Hindryckx, Jairath, & D'Haens, 2016; Podolsky, 2002). Genome-wide association studies of IBD patients have identified several susceptibility loci associated with dysregulated innate and adaptive immune responses (Duerr et al., 2006; Ogura et al., 2001; Xavier & Podolsky, 2007).

IBD encompasses a heterogeneous group of conditions that are distinguished by specific histological hallmarks of tissue damage. Crohn's disease (CD) is characterized by transmural inflammation, intestinal crypt abscesses, and granuloma formation, affecting any part of the whole gastrointestinal tract (Baumgart & Sandborn, 2012). Ulcerative colitis (UC) is typically associated with inflammation and ulceration that is limited to the mucosal layer of the colon only (Hanauer, 2004). The peak age of presentation with CD or UC is in adolescents or young adults. In the cases of colitis where histologically differentiating UC from CD is not clear cut, the term indeterminate colitis (IC) is assigned (Geboes et al., 2008). Very early onset IBD (VEOIBD) describes IBD that presents in children under the age of 6, when genetic susceptibility is likely to play a more important role (Benchimol et al., 2014). Microscopic colitis (MC) is an umbrella term for a subgroup of colitis with distinct histopathological patterns but no macroscopically evident abnormalities on endoscopy, which mostly affects elderly population (Langner et al., 2015). MC includes collagenous colitis (CC) and lymphocytic colitis (LC). LC features an increase in the number of surface intraepithelial lymphocytes, while CC classically demonstrates a thickened collagen band in the subepithelium (Langner et al., 2015).

The focus of research in IBD thus far has been directed toward investigating the cellular responses to inflammation. Hence, new therapies have focused on targeting the inflammatory pathways with limited efficacy (Magro et al., 2014). However, increasing evidence supports the important role of ECM in IBD pathogenesis. An imbalanced control of ECM synthesis vs degradation is believed to be a fundamental defect in IBD. Recent data further suggest that dynamic ECM remodeling also contributes to the chronicity of immune responses in IBD (Petrey & de la Motte, 2017; Shimshoni, Yablecovitch, Baram, Dotan, & Sagi, 2015). In this section, we highlight and discuss the key ECM compositional changes and remodeling associated with IBD pathogenesis.

3.2 ECM Composition in IBD

The intestinal ECM is essential for maintenance of the epithelial barrier integrity. An interruption of ECM conformation may promote infiltration of inflammatory cells into the lamina propria and submucosal space of the intestinal wall. Here, we provide an overview of various compositional changes in the ECM associated with IBD that have been previously described.

Excessive collagen deposition is a key histopathological hallmark associated with intestinal fibrosis in severe cases of IBD (Rieder & Fiocchi, 2009). For instance, collagens I, III, and V are upregulated in CD, causing tissue thickening across the mucosa, submucosa, and muscle layers of the intestinal wall (Pucilowska, Williams, & Lund, 2000; Stallmach, Schuppan, Riese, Matthes, & Riecken, 1992). Fibrosis of the intestinal wall causes stenosis of the lumen that commonly requires surgical resection. In contrast to CD, fibrosis in UC is limited to the muscularis mucosa, while expression of collagens IV and V is increased in the basement membrane (Pucilowska et al., 2000; Schmehl, Florian, Jacobasch, Salomon, & Korber, 2000; Stallmach et al., 1992). These matrix changes correlate positively with disease severity (Schmehl et al., 2000). Mesenchymal cells of the lamina propria are also altered by the fibrogenic phenotype with increased abundance of myofibroblasts, leading to enhanced ability to trigger ECM synthesis (Gelbmann et al., 1999). Excess collagen deposition is also seen in collagenous colitis coupled with increased TGF-β, VEGF, and FGF expression (Stahle-Backdahl et al., 2000; Taha et al., 2003). A recent study showed that culturing human colonic fibroblasts on CD-like stiff matrices developed fibrogenic phenotype via mechanotransduction signaling (Johnson et al., 2013). The data suggested that intestinal fibrosis may be self-propagating and independent of inflammation.

Laminin deposition is increased across all layers of the intestine in CD (Koutroubakis et al., 2003; Simon-Assmann et al., 1998). Mast cells were found to colocalize with laminin in CD strictures, suggesting their role in laminin deposition and stricture development (Gelbmann et al., 1999). However, it is unclear if the increased laminin deposition is mediated by mast cells alone or in combination with other mesenchymal cell populations in the ECM niche, such as smooth muscle cells, to promote stricture formation (Gelbmann et al., 1999). Inflamed segments of CD demonstrate increased expression of laminin 111 ($\alpha1\beta1\gamma1$) and laminin 332 ($\alpha3\beta3\gamma2$) and decreased expression of laminin 211 ($\alpha2\beta1\gamma1$) in the intestinal crypts, while the $\alpha1$ subunit was also upregulated in the crypts of uninflamed CD tissues (Bouatrouss, Herring-Gillam, Gosselin, Poisson, & Beaulieu, 2000).

Fibronectin has been shown to regulate fibroblast cell behavior in IBD by modulating the expression of TGF-β (Leeb et al., 2004, 2003). Increased fibronectin deposition is associated with excessive TGF-β secretion, which, in turn, promotes fibroblast migration, proliferation, and activation into myofibroblasts. Furthermore, the binding of fibronectin to TNF-α has also been shown to promote chemotaxis and expression of the ECM remodeling enzyme MMP-9 in monocytes (Vaday et al., 2000).

Hyaluronan (HA) deposition was increased in inflamed intestine and has been demonstrated to precede the inflammatory infiltration in IBD (de la Motte, 2011; Kessler et al., 2008). Fragmented HA molecules can induce damage by provoking proinflammatory signaling cascades via Toll-like receptor (TLR) 2 and 4 (Riehl, Foster, & Stenson, 2012; Scheibner et al., 2006). HA also has a well-documented role in propagating fibrosis by modulating proliferation and differentiation of fibroblasts to myofibroblasts (Simpson et al., 2009; Webber, Jenkins, Meran, Phillips, & Steadman, 2009). In contrast, intact but not fragmented HA chains promote regulatory T cell differentiation and contribute to the support of immune tolerance (Bollyky et al., 2009).

Lumican modulates host sensing of bacterial lipopolysaccharides by TLR4 and neutrophil chemotaxis via integrins (Lohr et al., 2012). Lumican regulates intestinal homeostasis by aiding innate immune and inflammatory responses, while deletion of lumican exhibits an attenuated immune response to chemically induced colitis, leading to increased morbidity and tissue damage (Lohr et al., 2012).

Intestinal epithelial and endothelial barrier dysfunction is another key pathological feature of IBD. The intestinal microvasculature permeates the lamina propria to convey important signals to the ISC niche. Alteration of the ECM allows increased permeability of the vascular endothelium, which enables circulating lymphocytes to be delivered to the damaged intestinal tissue for wound healing and repair (Hatoum, Heidemann, & Binion, 2006). Dysregulation of ECM modification can lead to excessive permeability, which promotes extravasation of leukocytes, platelets, and proinflammatory cytokines (Binion et al., 1998). In IBD tissues, cell adhesion and ECM molecules are upregulated in endothelial cells to enhance the recruitment of leukocytes to the inflamed endothelium (Binion et al., 1998; Koizumi, King, Lobb, Benjamin, & Podolsky, 1992; Rijcken et al., 2007). This highlights the important link between ECM compositional changes and inflammatory response.

3.3 ECM Remodeling in IBD

ECM remodeling drives its compositional changes and hence is an integral part of IBD pathology. Similar to normal tissue homeostasis, the major players of ECM remodeling belong to the protease family MMPs, which have received the most research interest in the field of IBD (recently reviewed extensively in O'Sullivan, Gilmer, & Medina, 2015).

An imbalance between ECM synthesis and ECM degradation can clinically manifest in IBD as fibrostenotic disease (discussed earlier) or fistulating disease. ECM fragmentation by MMPs is significantly increased in inflamed IBD tissue, while inhibition of these proteolytic enzymes represents great therapeutic value (Baugh et al., 1999; O'Sullivan et al., 2015). This is complemented by studies showing that inhibitors of MMPs such as TIMP-3 are downregulated in inflamed Crohn's tissue. Likewise, depletion of TIMP-3 in mice promotes development of colitis (Monteleone et al., 2012). CD-related fistulas occur due to excessive ECM breakdown followed by an inappropriate epithelialization of an abnormal tract connecting the intestine with adjacent anatomical structures such as the skin, bladder, or other parts of the bowel. Fistula formation in CD has been associated with increased expression of MMP-3 and -9 (Kirkegaard, Hansen, Bruun, & Brynskov, 2004).

To date, synthetic MMP inhibitors have not been successful in treating or reversing intestinal fibrosis in clinical trials. A more recent study has identified another novel ECM remodeling protein KIAA1199 in IBD pathogenesis (Soroosh et al., 2016). KIAA1199 can degrade HA chains in the intestinal matrix, producing proinflammatory fragments to augment the chronicity of inflammation in an IL-6-driven mechanism (Soroosh et al., 2016). Current immunomodulatory therapies have little to no efficacy for fibrostenotic disease or fistulating disease. Better understanding of ECM remodeling will open exciting therapeutic avenue for targeting the tissue destructive fibrotic or fistulating processes in IBD.

4. ECM DYNAMICS IN CRC

4.1 CRC Microenvironment

CRC is the third leading cause of cancer death in the world (Torre et al., 2015). Numerous studies have identified driver mutations in several key signaling pathways involved in diverse stem cell biology such as proliferation and differentiation (Cancer Genome Atlas Network, 2012; Cristobal et al., 2017; Wood et al., 2007). However, the role of the tumor microenvironment in CRC progression is less characterized. Increasing evidence suggests the involvement of microenvironment-derived signals in tumor initiation and progression (Medema & Vermeulen, 2011). It is therefore important to consider the surrounding niche when studying CRC development as well as therapeutic intervention. Similar to the ISC niche discussed earlier, the tumor microenvironment consists of both ECM and cellular components that form a complex network of ECM proteins and cells surrounding the tumor. The ECM functions not only as structural scaffold for tissue organization but also as a rich source of biochemical and biomechanical cues to govern cell behavior (Hynes, 2009; Pickup, Mouw, & Weaver, 2014). In this section, we discuss the role of various components of the CRC microenvironment in tumor progression (Fig. 1C).

4.1.1 ECM Remodeling in CRC

The role of tumor microenvironment in CRC is not as well characterized as in other types of cancers such as breast cancer. The interstitial matrix of the ECM consists of fibrillary collagens and fibronectin among other macromolecules to provide structural framework as well as reservoir for growth factors and cytokines to bind (Egeblad, Nakasone, & Werb, 2010). TGF-β has been shown to trigger ECM synthesis by activating fibroblasts to differentiate into a secretory myofibroblast phenotype (Barrientos, Stojadinovic, Golinko, Brem, & Tomic-Canic, 2008). The fibrotic ECM is one of the most notable features in cancer microenvironment as a result of excessive ECM synthesis, where ECM is organized in relatively straight/parallel alignment (Malik, Lelkes, & Cukierman, 2015). Deposition of fibrous connective tissue such as collagen and fibronectin is usually a transient response in normal wound healing, which can persist in the event of repetitive or severe tissue injury (Wynn & Ramalingam, 2012). The irreversible fibrotic response in cancer is carried out by cancer-associated fibroblasts (CAFs) to form the bulk of interstitial ECM (Malik et al., 2015). The straight and organized architecture of the tumor ECM increases the matrix stiffness compared to normal tissues, which can, in turn, alter cell behavior.

Multiple enzymes have been shown to contribute to ECM remodeling. Fibroblast activation protein (FAP) is an ECM-modifying enzyme that can cause desmoplastic alignment of ECM (Kim, Watkinson, Varadan, & Anastassiou, 2010; Malik et al., 2015). Notably, FAP has been suggested as a CRC prognostic marker where its expression level is associated with tumor aggressiveness and poor patient outcome (Calon et al., 2015). High LOX expression has also been observed in advanced CRCs, while deletion of LOX in CRC led to reduced invasion (Baker et al., 2011). LOX activity has been shown to contribute to tissue stiffness and aligned collagen crosslinking structure rather than collagen deposition (Baker, Bird, Lang, Cox, & Erler, 2012). Such collagen modification of the matrix promotes cell proliferation via activation of integrin followed by focal adhesion kinase signaling for mechanotransduction (Baker et al., 2012; Tilghman & Parsons, 2008).

Alteration of tissue stiffness and architecture by ECM remodeling in cancer is a complex process involving both mechanical and cellular participants. Indeed, dynamic interplay between biochemical and mechanical factors is crucial for ECM remodeling (Malik et al., 2015). CAFs contribute to ECM deposition and remodeling, while their maintenance is dependent on stiffness of the matrix and the surrounding cytokines (Wynn & Ramalingam, 2012). In addition to fibrotic ECM remodeling, the biophysical properties of the ECM can also be modified to promote tumor growth and progression. The elasticity and porosity of ECM, shaped by a number of components such as MMPs and integrins, can directly influence the ability of cells to invade through the ECM (Wolf et al., 2013). The increased cell infiltration will, in turn, reinforce ECM remodeling. It has been suggested that reversing ECM remodeling in cancer, such as by reducing stiffness and alignment, may represent a more effective therapeutic approach for cancer rather than removing desmoplasia (Malik et al., 2015).

The proteolytic enzymes MMPs also play a crucial role in cancer progression. MMP-7 activity was found to be localized predominantly at the tumor-stromal boundary in a CRC mouse model (Scherer, McIntyre, & Matrisian, 2008). MMPs can be activated by increased mechanical force such as stiffer ECM caused by increased collagen production (Seliktar, Nerem, & Galis, 2001). MMP-1 is one of the ECM remodeling enzymes that degrades fibrillar collagens and has been implicated in cell invasion in vivo (Kalluri, 2016; Malik et al., 2015). MMP-3 has been shown to cleave E-cadherin and promote epithelial–mesenchymal transition (EMT) of the cancer cells (Lochter et al., 1997). Overexpression of MMP-11 in CRC is associated with poor patient prognosis, while its expression is found in mesenchymal cells rather than malignant epithelial cells (Porte et al., 1995). It should be noted that the MMP activity is regulated by MMP inhibitors such as TIMPs that are often found in tumor sites (Deryugina & Quigley, 2006). MMPs therefore play a pivotal role in CRC progression via ECM remodeling.

4.1.2 Mesenchymal Cells in CRC Niche

Similar to normal ISC niche, the tumor microenvironment consists of various cell types such as fibroblasts, immune cells, and endothelial cells. In particular, CAFs and immune cells play an indispensable role in tumor progression by secreting growth factors/chemokines and participating in matrix remodeling (Erez, Truitt, Olson, & Hanahan, 2010; Mantovani, Romero, Palucka, & Marincola, 2008).

Stromal cell-dependent signaling plays a critical role in tumor progression. CAFs have been shown to enhance proliferation and tumorigenesis in cancer cells when compared to normal fibroblasts (Kalluri, 2016). As discussed earlier, activation of Wnt signaling and suppression of BMP signaling at the niche are key for ISC homeostasis. Likewise, these signaling pathways are equally important in CRC microenvironment. BMP antagonist Gremlin-1 is strongly expressed by CAFs in many cancer types including CRCs (Sneddon et al., 2006). Aberrant Wnt activation is frequently observed in CRC, where majority of CRCs harbor mutations in Wnt pathway components (Humphries & Wright, 2008; Reya & Clevers, 2005). Cellular heterogeneity of Wnt activity is observed in CRCs with elevated Wnt signaling in tumor cells that are in close proximity to CAFs, suggesting that the CRC cells may be in part regulated by the tumor microenvironment to maintain high Wnt activity (Vermeulen et al., 2010; Voloshanenko et al., 2013). Metastatic CRC is associated with loss of basement membrane, exposing cancer cells at the invasive edge to a different microenvironment (Spaderna et al., 2006). This results in EMT, which is often accompanied by enhanced Wnt signaling and loss of differentiation (Brabletz & Kirchner, 2005). Mutations in Smad4, a downstream component of the TGF-β pathway, was further shown to promote stromal cells hyperproliferation in an intestinal tumor model (Takaku et al., 1998). On the other hand, gene signature induced by TGF-β signaling in CAFs is associated with poor prognosis in CRCs (Calon et al., 2015). It is interesting to note that activation of the transcription factor YAP in CAFs has been shown to promote ECM stiffness, which, in turn, further enhances YAP activation in a feedforward mechanism to maintain CAF phenotype (Calvo et al., 2013). More work on the paracrine signaling between the stromal microenvironment and tumor cells will be needed to better understand the role of CAFs in tumor progression.

Notably, CAFs are found to promote cancer cell proliferation and invasion by carrying out ECM synthesis and remodeling via proteinases such as MMPs and TIMPs (Egeblad et al., 2010; Kalluri, 2016). Indeed, cancer cells are believed to indirectly promote their own growth by recruiting stromal cells, which are the major source of production of these proteases (Kessenbrock, Plaks, & Werb, 2010). For example, CRC cells have been demonstrated to promote CAF infiltration and proliferation by increased expression of TIMP-1 that binds to its receptor CD63 on CAFs for protumor growth factors and cytokines secretion (Gong et al., 2013). Activation of CAFs by TGF-β1 triggers ECM deposition such as collagen, MMPs, and TIMPs (Hawinkels et al., 2012). Osteopontin (OPN) is secreted by cancer cells to induce TGF-β1 production and subsequent activation of CAFs in breast cancer model, while its expression is also found upregulated in CRC (Fedarko, Jain, Karadag, Van Eman, & Fisher, 2001; Weber et al., 2015). Interestingly, ECM proteins also contain cryptic structures similar to cytokines that can be exposed upon proteolysis with distinct functions from the full-length protein (Monboisse et al., 2013). The cleaved ECM components have been implicated in recruiting immune cells, another key stromal population that comprises the tumor microenvironment (Houghton et al., 2006; Weathington et al., 2006). Although their role in CRC has not been explored, ECM fragments are also shown to be capable of stimulating cancer cells and convey tumorigenic effect (Blood & Zetter, 1989; Pocza, Suli-Vargha, Darvas, & Falus, 2008).

CAFs have also been implicated in regulating tumor immunity directly by secreting immunomodulatory cytokines and chemokines. It has been shown that CAFs express proinflammatory genes when induced by IL-1β secreted from the surrounding immune cells (Erez et al., 2010). Moreover, TGF-β1-activated CAFs are found to secrete IL-11 to promote CRC cell survival and their colonization (Calon et al., 2012). Inflammation has been observed following common driver mutations such as oncogenic RAS to facilitate cancer progression (Sparmann & Bar-Sagi, 2004). Immune cells attracted to the site of inflammation produce a variety of growth factors and cytokines including TNF-α and TGF-β that can promote tumor initiation (Grivennikov, Greten, & Karin, 2010). Indeed, chronic inflammation of the gut such as in IBD patients is often predisposed to CRC (Itzkowitz & Yio, 2004). Nuclear factor-κB (NF-κB) signaling plays a crucial role in mediating inflammation (Karin, 2006). In a carcinogen-induced CRC model, cancer initiation and growth were promoted by NF-κB activity in epithelial and myeloid cells, respectively, indicating the close link between inflammatory response and tumorigenesis (Greten et al., 2004).

Innate immune cells are mainly responsible for inflammatory response in cancer. Several myeloid cells such as macrophages and neutrophils responsible for inflammatory reactions in normal tissue regeneration are also involved in tumorigenesis. Macrophages promote angiogenesis in wound healing and have two activation states, classical M1 and alternative M2, polarized by cytokine milieu secreted via T helper cells or microbial products (Mantovani et al., 2008). M1 macrophages act against intracellular pathogens and cancer cells, while M2 macrophages arise in defense against parasites and for relevant tissue remodeling (Mantovani et al., 2008). Unlike M1 and M2, the tumor-associated macrophages (TAMs) have unique identities to promote tumor growth and metastasis by secreting ECM remodeling enzymes, angiogenic factors, and growth factors (Condeelis & Pollard, 2006; Egeblad et al., 2010; Pollard, 2009). Inhibition of NF-κb signaling in TAMs has been shown to cause cytotoxicity to tumor cells by abrogating the immunosuppressive function and driving them toward M1-like phenotype (Hagemann et al., 2008). Similarly, neutrophils also have two distinct tumor-associated subtypes based on their effect on tumor growth, where TGF-β signaling drives the polarization toward protumorigenic subtype by secreting ECM remodeling enzymes and angiogenic factors (Fridlender et al., 2009).

4.2 ECM Remodeling for Invasion and Metastasis

Metastasis is a complex process in which cancer cells disseminate from the primary tumor and colonize another body part(s). ECM remodeling is required to establish local microenvironment at both primary and metastatic sites for cancer cells to invade and colonize. At the primary tumor site, ECM is remodeled to permit migration of cancer cells via EMT across the basement membrane and muscularis mucosa, where tumor cells are exposed to the stromal microenvironment (Brabletz et al., 2005; Spaderna et al., 2006). At the metastatic site of a distant organ, ECM niche is also required to allow circulating cancer cells to colonize and proliferate (Psaila & Lyden, 2009). In this section, we focus on the discussion of the ECM composition and remodeling at the primary site that facilitates tumor cell invasion.

Several ECM components have been reported to be dysregulated in invasive CRC. Type I collagen plays a crucial role to induce an ISC-like phenotype reminiscent of EMT for tumor cells to invade, and is found highly expressed in tumors with infiltrative growth (Brabletz et al., 2004; Kirkland, 2009; Oku et al., 2008). Similarly, the glycoprotein tenascin C (TNC) is also found to be highly expressed in the collective tissues around the invading tumor edge (Jones, 2001). Overexpression of TNC in some CRC may enhance tumor cell migration due to its previously described antiadhesive property (Swindle et al., 2001). Expression of Galectin-3, among other matrisome genes, is also found to be associated with metastasis of CRCs to liver (Naba et al., 2014; Nakamura et al., 1999). On the other hand, the β-galactoside-binding protein galectin-8 has a suppressive role in cell migration and is downregulated in malignant CRCs (Nagy et al., 2002).

Alterations of ECM peptidases are also observed during tumor invasion. CAF-derived MMP-2 has been shown to degrade the basement membrane protein, type IV collagen, to permit initial invasion into the lamina propria (Taniwaki et al., 2007). The membrane-type 1 MMP (MT1-MMP/MMP-14) is responsible in activating pro-MMP2, and is found upregulated in CRC cells at the invading (Ohtani, Motohashi, Sato, Seiki, & Nagura, 1996). Furthermore, computational analysis of public datasets of several cancers including CRC reveals a metastasis-associated gene expression signature, which contains other proteolytic degradation and ECM remodeling genes such as MMP-2/11, LOX, and FAP that are believed to facilitate continued invasion into the stroma (Kim et al., 2010). TGF-β signaling is also found activated in aggressive CRC subtypes (Cancer Genome Atlas Network, 2012; Guinney et al., 2015; Kim et al., 2010). Interestingly, treatment of CAFs with TGF-β has been shown to induce MMP-2 and -9 activities and cause higher rates of CRC metastasis to lung and liver, suggesting a metastatic-role of TGF-β signaling via ECM remodeling (Calon et al., 2012; Hawinkels et al., 2012). Although the exact mechanism of CRC invasion remains elusive, these findings suggest that ECM remodeling activity plays a key role in tumor progression.

In addition to alterations of ECM composition and remodeling enzymes, the arrangement of collagen fibers themselves has also been associated with metastasis in many cancer types. Collagen fibers in the surrounding stroma are increasingly aligned as tumor progresses, which differ from the random arrangement observed in normal tissue (Malik et al., 2015). This rearrangement is of particular relevance to epithelial invasion where the fibers act as highway-like structure to guide cancer cell migration (Provenzano et al., 2006). In summary, ECM remodeling in the tumor microenvironment contributes a significant role during tumor invasion and is closely associated with cancer progression and prognosis.

4.3 CRC Therapeutic Strategies Targeting Tumor Microenvironment

Current cancer drug development targets mainly dysregulated signaling pathways in cancer cells. On the other hand, increasing evidence supports the importance of stromal cells and ECM for cancer progression, suggesting that therapeutic strategies targeting the tumor microenvironment may be exploited. Potential targets of ECM remodeling include not only ECM peptidases but also cell surface receptors that interact with ECM, as well as cancer-associated stromal cells (Bonnans et al., 2014).

MMP inhibitors were initially thought to be promising for dysregulation of ECM peptidases in CRC, yet subsequent research and clinical trials failed to prove efficacy due to their poor specificity and severe toxicity (Coussens, Fingleton, & Matrisian, 2002; Drummond et al., 1999). As a result, a new strategy of selective MMP inhibitor was developed using specific monoclonal antibodies to target individual MMPs (Devy & Dransfield, 2011). The LOXL-2-specific inhibitor, Simtuzumab, has been used in murine fibrosis models and is now in clinical trials for cancer (Barry-Hamilton et al., 2010). On the other hand, the EGF receptor inhibitor, Cetuximab, has been shown to trigger fibronectin synthesis that reduces the cytotoxic effects of chemotherapy (Eke, Storch, Krause, & Cordes, 2013). Of note, CAF activity often increases intratumoral interstitial fluid pressure and matrix stiffness, which subsequently may interfere with anticancer drug delivery and uptake (Heldin, Rubin, Pietras, & Ostman, 2004; Junttila & de Sauvage, 2013).

An alternative approach is to target cancer-associated immune cells. It has been shown that integrin-mediated interactions with the ECM can recruit macrophages to the tumor and trigger angiogenesis to facilitate invasion (Jin, Su, Garmy-Susini, Kleeman, & Varner, 2006). A multikinase inhibitor, Regorafenib, blocks angiogenesis by targeting TAMs that express receptor tyrosine kinase, while this drug is highly toxic over prolonged treatment (Abou-Elkacem et al., 2013; Argilés et al., 2015; Murphy et al., 2016). Despite the promising therapeutic development against tumor microenvironment, continued research and more clinical trial data are still required to monitor the drug efficacy and safety. Better understanding of the pro-tumorigenic roles of the ECM and stromal cells in the tumor microenvironment will help develop effective cancer preventive and therapeutic strategies.

5. TISSUE ENGINEERING

5.1 Intestinal Tissue Engineering

Intestinal failure caused by diseases such as short bowel syndrome (SBS) ultimately requires transplantation. Persistent clinical challenges such as limited donor organs, graft-host rejection complications as well as long-term immunosuppression indicate an urgent need to identify new treatments. Recent advances in the field of tissue engineering suggest an alternative novel therapy for SBS patients, for which reliable curative treatments remain suboptimal (Cavicchi, Beau, Crenn, Degott, & Messing, 2000; Middleton & Jamieson, 2005; Vanderhoof & Langnas, 1997).

Tissue engineering requires the combination of biocompatible scaffolds with stem cells to create a regenerated tissue construct for orthotopic transplantation. Scaffolds are essential to provide structural support as well as biochemical cues to permit stem cells repopulation. It is therefore important that scaffolds are biomimetic and resemble the ECM microenvironment in vivo. Scaffolds can be either synthetic or biological. Synthetic scaffolds are generally made of biodegradable polymers that are molded or printed into the desired organ shape. However, limited success has been achieved using these synthetic scaffolds for tissue engineering, possibly due to the lack of biological cues. Biological scaffolds, on the other hand, derived from the ECM of an organ via decellularization are inherently more biomimetic than synthetic scaffolds. Decellularization is achieved by perfusing an organ with detergent and enzyme solutions to remove cellular material and other immunogens, while retaining the native ECM composition and structure, thus representing an ideal biological scaffold to support stem cell-based tissue engineering. In fact, a stem cell-based tissue engineered trachea using decellularized scaffold has already been successfully transplanted to a pediatric patient with congenital tracheal stenosis, where a functional airway was seen at 2-year follow-up (Elliott et al., 2012).

Decellularization-based tissue engineering in heart, lung, and kidney has been previously reported in small animals, while successful intestinal tissue engineering (ITE) has yet to be revealed (Ott et al., 2010, 2008; Petersen et al., 2010; Song et al., 2013). Decellularization protocol of small intestine has been demonstrated in rat that showed preservation of the native architecture and ECM components such as collagen and GAG (Totonelli et al., 2012). Recent progress on ITE has been made by seeding intestinal organoids and fibroblasts on either synthetic or decellularized scaffolds, while the focus was mainly on optimizing culture conditions to promote epithelial differentiation rather than complete reconstruction of the whole intestine (Schweinlin et al., 2016; Wang et al., 2017). To date, there has been no attempt to recreate the musculature or vasculature of the intestine, indicating that considerable future work is still required in this area. As discussed earlier, the ISC niche includes many components other than fibroblasts, including endothelial cells and neural cells. Isolation and transplantation of functional enteric neurons from pediatric patients have been described, while others have investigated regeneration of muscle layers in esophagus (Metzger, Caldwell, Barlow, Burns, & Thapar, 2009; Poghosyan et al., 2015). Similarly, decellularized skin has been revascularized using microvascular endothelial cells (Groeber et al., 2016). The next milestone of ITE will be to combine all these approaches for complete reconstruction of engineered intestine for clinical transplantation (Fig. 2).

5.2 CRC Modeling

Alternatively, tissue engineering of intestine can be applied for modeling of CRC in vitro. Recent advances in ex vivo organoid culture derived from primary human CRC tissues offer excellent model for drug testing (van de Wetering et al., 2015). However, the lack of tumor microenvironment and tissue architecture in the organoid culture raises concern about its biomimicry. Potential 3D reconstruction of CRC with associated tumor microenvironment may offer an improved system for disease modeling and drug screening.

5.2.1 Overview of Current Status of CRC Tissue-Engineered Models

The vast majority of the current tissue-engineered CRC models rely mostly on the use of established CRC cell lines, often in coculture with other stromal cell types such as fibroblasts and endothelial cells in natural or synthetic scaffolds (Nyga, Cheema, & Loizidou, 2011). For instance, a 3D biomimetic CRC model has been established by embedding CRC cell lines (HT29 or HCT116) within dense collagen hydrogels to form "artificial cancer masses" that were surrounded by a loose collagenous matrix containing fibroblasts and endothelial cells (Magdeldin et al., 2017; Nyga, Loizidou, Emberton, & Cheema, 2013). Migration of the epithelial cells from the cancer mass to invade the surrounding matrix was observed using this system, where presence of laminin enhanced tumor invasion. In another system, a microfluidic platform was used to reconstruct 3D microtumors with vascular networks by seeding cancer cell lines, endothelial cells, and fibroblasts in fibrinogen/thrombin polymerized gels under dynamic flow (Sobrino et al., 2016). The system was then used to test the efficacy of drugs targeting vasculature. However, despite the inclusion of some aspects of the stromal microenvironment, these models still lack the physical complexity such as tissue architecture, tissue-specific ECM composition, and mechanical properties to recapitulate CRC development in vivo.

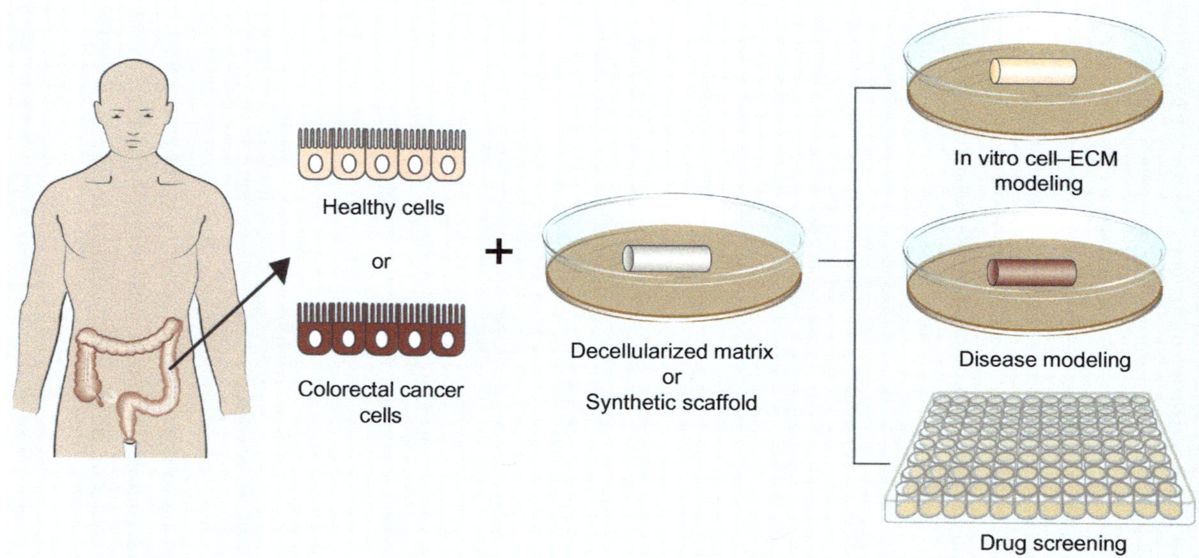

Fig. 2 Potential applications of tissue-engineered intestine. Primary normal or cancerous intestinal cells could be isolated from individuals. These cells could be expanded in vitro to repopulate the decellularized matrices or synthetic scaffolds to reconstruct neo-intestinal tissue in vitro. The tissue-engineered intestine could then be used for autologous transplantation or disease modeling. The presence of tumor microenvironment in the engineered CRC construct represents a promising biomimetic cancer model for drug screening, which may potentially be used for personalized therapies for CRC.

Another 3D tumor model was developed by seeding cancer cell lines with primary fibroblasts and/or endothelial cells onto decellularized porcine small intestinal scaffolds with biomimetic dynamic flow (Moll et al., 2013; Nietzer et al., 2016). This has greatly improved 3D tumor modeling with better structural and biological mimicry. Furthermore, administration of the chemotherapy drug 5-FU was further shown to be effective only in the 3D model but not in 2D system, highlighting the importance of complete reconstruction of the tumor microenvironment. More recently, Chen and colleagues described yet another organotypic CRC model by repopulating decellularized human colon matrix with primary colonic epithelial cells derived from organoids together with myofibroblasts and endothelial cells (Chen et al., 2016). Seeding of colonic epithelial cells carrying various driver mutations onto the system further recapitulated the in vivo malignant phenotype where CRC cells invaded through the muscularis mucosa into the submucosa. These 3D engineered CRC models highlight the crucial role of both stromal cells and ECM in the tumor microenvironment.

5.2.2 Tissue-Engineered Intestine as a Model for Drug Testing

Cancers are mostly caused by accumulation of multiple oncogenic mutations. In addition to the traditional chemotherapy and radiation therapy, many new drugs have been developed in the past decades to target specific gene mutations or pathways involved. Nevertheless, the variation of drug effectiveness between individuals is common, where patients can develop intrinsic or adaptive drug resistance. For example, the drug (PLX4032) targeting *BRAF* (*V600E*) mutation is found insensitive in CRCs carrying the same mutation due to the feedback activation of EGFR-mediated MAPK signaling (Corcoran et al., 2012; Prahallad et al., 2012). On the other hand, treatment of PLX4032 is highly effectively in *BRAF* (*V600E*)-mutated melanoma, while acquired resistance is observed in some patients with upregulated EGFR expression (Sun et al., 2014). A new strategy of personalized medicine is urgently needed to identify patient group with specific genetic alteration that responds to a particular drug or combination therapy. This requires a representative physiological cancer model for drug development and testing.

Most preclinical studies in drug development reply on either 2D cancer cell culture or animal models such as patient-derived xenograft (PDX) models, while the rate of successful translation to clinical trials for cancer therapy remains dismally low (Mak, Evaniew, & Ghert, 2014; Perel et al., 2007).

As discussed, 2D culture condition lacks physiological relevance for disease modeling. Although specific disease processes can be tested in animal models, comprehensive human disease conditions and progression may not be recapitulated due to the physiological variation between mouse and human. In addition, the high cost and time-consuming strategy of PDX model further limits its application on therapeutic development.

The tumor-derived organoid technology has opened up exciting avenue for disease modeling and drug screening. Whole genome sequencing and proteomic analysis further demonstrate that the identities and profiles of individual CRC patients are conserved in the established organoid culture compare with the primary tumor (Cristobal et al., 2017; van de Wetering et al., 2015). Importantly, these tumor organoids render possible high-throughput drug screening and personalized medicine for individual patients. However, it is unclear if the lack of the native ECM and stromal cell microenvironment in the culture can faithfully recapitulate CRC in vivo. The presence of excessive growth factors in the culture media further argues its physiological accuracy for drug response. The engineered 3D cancer model appears to be the new approach for disease modeling. Reconstruction of CRC in vitro using decellularized ECM scaffold in combination with primary tumor organoids and cancer stromal cells will provide a biomimetic cancer model with tumor microenvironment that recapitulates CRC in vivo (Fig. 2). Growing evidence, as discussed earlier, highlights the important role of ECM in both normal ISC and CRC development. The cancer engineering will improve the current cancer models for drug testing by including both stromal cells and ECM tumor microenvironment. More work is needed to further optimize the engineered CRC model to allow high-throughput drug screening for new generation of personalized medicine.

6. CONCLUSION

Stem cell niche plays an indispensable role in tissue homeostasis, regeneration, and disease. Current view of ISC niche focuses mainly on mesenchymal cells and Paneth cells, while the importance of ECM is less revealed. In this chapter, we provided an overview of the ECM composition and remodeling in intestinal homeostasis, inflammation, and cancer. In addition to the structural support, ECM provides both biophysical and biochemical cues to regulate cell behavior via cell–ECM interaction and the availability of growth factors. The epithelial and stromal cells produce

ECM remodeling enzymes to modify the ECM microenvironment, which, in turn, regulates ISC fate decision. Such feedback regulation between cells and ECM is crucial in shaping the microenvironment for tissue homeostasis, while disruption of the process often leads to diseases such as IBD and cancer. ECM remodeling enzymes are the central players in the ISC niche to support tissue homeostasis, wound healing, and tumor growth in different contexts. ECM components therefore lend promises to CRC drug development as an alternative therapeutic strategy. On the other hand, the natural ECM obtained from decellularized intestine offers an excellent biological scaffold for tissue engineering. Ex vivo reconstruction of CRC using such tissue engineering technology further allows incorporation of the tumor microenvironment to cancer cells, representing a promising biomimetic tool for disease modeling and drug testing.

ACKNOWLEDGMENTS

This review is a snapshot of ECM remodeling in intestinal homeostasis and disease at the current status, and we apologize to many colleagues whose work could not be cited here due to space limitation. This work was supported by the Francis Crick Institute which receives its core funding from Cancer Research UK (FC001105), the UK Medical Research Council (FC001105), and the Wellcome Trust (FC001105); European Union's Horizon 2020 research and innovation programme (668294).

REFERENCES

Abou-Elkacem, L., Arns, S., Brix, G., Gremse, F., Zopf, D., Kiessling, F., et al. (2013). Regorafenib inhibits growth, angiogenesis, and metastasis in a highly aggressive, orthotopic colon cancer model. *Molecular Cancer Therapeutics*, *12*(7), 1322–1331.

Argilés, G., Saunders, M. P., Rivera, F., Sobrero, A., Benson, A., III, Guillén Ponce, C., et al. (2015). Regorafenib plus modified FOLFOX6 as first-line treatment of metastatic colorectal cancer: A phase II trial. *European Journal of Cancer*, *51*(8), 942–949.

Baker, A.-M., Bird, D., Lang, G., Cox, T. R., & Erler, J. T. (2012). Lysyl oxidase enzymatic function increases stiffness to drive colorectal cancer progression through FAK. *Oncogene*, *32*(14), 1863–1868.

Baker, A.-M., Cox, T. R., Bird, D., Lang, G., Murray, G. I., Sun, X.-F., et al. (2011). The role of lysyl oxidase in SRC-dependent proliferation and metastasis of colorectal cancer. *Journal of the National Cancer Institute*, *103*(5), 407–424.

Bamba, S., Andoh, A., Yasui, H., Araki, Y., Bamba, T., & Fujiyama, Y. (2003). Matrix metalloproteinase-3 secretion from human colonic subepithelial myofibroblasts: Role of interleukin-17. *Journal of Gastroenterology*, *38*(6), 548–554.

Barker, N., van Es, J. H., Kuipers, J., Kujala, P., van den Born, M., Cozijnsen, M., et al. (2007). Identification of stem cells in small intestine and colon by marker gene Lgr5. *Nature*, *449*(7165), 1003–1007.

Barrientos, S., Stojadinovic, O., Golinko, M. S., Brem, H., & Tomic-Canic, M. (2008). Growth factors and cytokines in wound healing. *Wound Repair and Regeneration*, *16*(5), 585–601.

Barry-Hamilton, V., Spangler, R., Marshall, D., McCauley, S., Rodriguez, H. M., Oyasu, M., et al. (2010). Allosteric inhibition of lysyl oxidase-like-2 impedes the development of a pathologic microenvironment. *Nature Medicine, 16*(9), 1009–1017.

Basora, N., Vachon, P. H., Herring-Gillam, F. E., Perreault, N., & Beaulieu, J. F. (1997). Relation between integrin alpha7Bbeta1 expression in human intestinal cells and enterocytic differentiation. *Gastroenterology, 113*(5), 1510–1521.

Basson, M. D., Turowski, G., & Emenaker, N. J. (1996). Regulation of human (Caco-2) intestinal epithelial cell differentiation by extracellular matrix proteins. *Experimental Cell Research, 225*(2), 301–305.

Baugh, M. D., Perry, M. J., Hollander, A. P., Davies, D. R., Cross, S. S., Lobo, A. J., et al. (1999). Matrix metalloproteinase levels are elevated in inflammatory bowel disease. *Gastroenterology, 117*(4), 814–822.

Baumgart, D. C., & Sandborn, W. J. (2012). Crohn's disease. *Lancet, 380*(9853), 1590–1605.

Beaulieu, J. F. (1992). Differential expression of the VLA family of integrins along the crypt-villus axis in the human small intestine. *Journal of Cell Science, 102*(Pt. 3), 427–436.

Beaulieu, J. F., & Vachon, P. H. (1994). Reciprocal expression of laminin A-chain isoforms along the crypt-villus axis in the human small intestine. *Gastroenterology, 106*(4), 829–839.

Benchimol, E. I., Mack, D. R., Nguyen, G. C., Snapper, S. B., Li, W., Mojaverian, N., et al. (2014). Incidence, outcomes, and health services burden of very early onset inflammatory bowel disease. *Gastroenterology, 147*(4), 803–813.e807; quiz e814–805.

Bhanja, P., Saha, S., Kabarriti, R., Liu, L., Roy-Chowdhury, N., Roy-Chowdhury, J., et al. (2009). Protective role of R-spondin1, an intestinal stem cell growth factor, against radiation-induced gastrointestinal syndrome in mice. *PLoS One, 4*(11), e8014.

Binion, D. G., West, G. A., Volk, E. E., Drazba, J. A., Ziats, N. P., Petras, R. E., et al. (1998). Acquired increase in leucocyte binding by intestinal microvascular endothelium in inflammatory bowel disease. *Lancet, 352*(9142), 1742–1746.

Blood, C. H., & Zetter, B. R. (1989). Membrane-bound protein kinase c modulates receptor affinity and chemotactic responsiveness of lewis lung carcinoma sublines to an elastin-derived peptide. *The Journal of Biological Chemistry, 264*(18), 10614–10620.

Bollyky, P. L., Falk, B. A., Wu, R. P., Buckner, J. H., Wight, T. N., & Nepom, G. T. (2009). Intact extracellular matrix and the maintenance of immune tolerance: High molecular weight hyaluronan promotes persistence of induced cd4+cd25+ regulatory t cells. *Journal of Leukocyte Biology, 86*(3), 567–572.

Bonnans, C., Chou, J., & Werb, Z. (2014). Remodelling the extracellular matrix in development and disease. *Nature Reviews. Molecular Cell Biology, 15*(12), 786–801.

Bouatrouss, Y., Herring-Gillam, F. E., Gosselin, J., Poisson, J., & Beaulieu, J. F. (2000). Altered expression of laminins in Crohn's disease small intestinal mucosa. *The American Journal of Pathology, 156*(1), 45–50.

Brabletz, T., Hlubek, F., Spaderna, S., Schmalhofer, O., Hiendlmeyer, E., Jung, A., et al. (2005). Invasion and metastasis in colorectal cancer: Epithelial-mesenchymal transition, mesenchymal-epithelial transition, stem cells and ß-catenin. *Cells, Tissues, Organs, 179*, 56–65.

Brabletz, T., Spaderna, S., Kolb, J., Hlubek, F., Faller, G., Bruns, C. J., et al. (2004). Down-regulation of the homeodomain factor Cdx2 in colorectal cancer by collagen type I: An active role for the tumor environment in malignant tumor progression. *Cancer Research, 64*(19), 6973–6977.

Breynaert, C., de Bruyn, M., Arijs, I., Cremer, J., Martens, E., Van Lommel, L., et al. (2016). Genetic deletion of tissue inhibitor of metalloproteinase-1/TIMP-1 alters inflammation and attenuates Fibrosis in Dextran Sodium Sulphate-induced Murine Models of Colitis. *Journal of Crohn's & Colitis, 10*(11), 1336–1350.

Calon, A., Espinet, E., Palomo-Ponce, S., Tauriello, D. V., Iglesias, M., Cespedes, M. V., et al. (2012). Dependency of colorectal cancer on a TGF-beta-driven program in stromal cells for metastasis initiation. *Cancer Cell*, *22*(5), 571–584.

Calon, A., Lonardo, E., Berenguer-Llergo, A., Espinet, E., Hernando-Momblona, X., Iglesias, M., et al. (2015). Stromal gene expression defines poor-prognosis subtypes in colorectal cancer. *Nature Publishing Group*, *47*(4), 320–329.

Calvo, F., Ege, N., Grande-Garcia, A., Hooper, S., Jenkins, R. P., Chaudhry, S. I., et al. (2013). Mechanotransduction and YAP-dependent matrix remodelling is required for the generation and maintenance of cancer-associated fibroblasts. *Nature Cell Biology*, *15*(6), 637–646.

Cancer Genome Atlas Network. (2012). Comprehensive molecular characterization of human colon and rectal cancer. *Nature*, *487*(7407), 330–337.

Cavicchi, M., Beau, P., Crenn, P., Degott, C., & Messing, B. (2000). Prevalence of liver disease and contributing factors in patients receiving home parenteral nutrition for permanent intestinal failure. *Annals of Internal Medicine*, *132*(7), 525–532.

Chen, P., Abacherli, L. E., Nadler, S. T., Wang, Y., Li, Q., & Parks, W. C. (2009). MMP7 shedding of syndecan-1 facilitates re-epithelialization by affecting alpha(2)beta(1) integrin activation. *PLoS One*, *4*(8), e6565.

Chen, H. J., Wei, Z., Sun, J., Bhattacharya, A., Savage, D. J., Serda, R., et al. (2016). A recellularized human colon model identifies cancer driver genes. *Nature Biotechnology*, *8*, 845–851.

Cheng, H., & Leblond, C. P. (1974a). Origin, differentiation and renewal of the four main epithelial cell types in the mouse small intestine. I. Columnar cell. *The American Journal of Anatomy*, *141*(4), 461–479.

Cheng, H., & Leblond, C. P. (1974b). Origin, differentiation and renewal of the four main epithelial cell types in the mouse small intestine. V. Unitarian theory of the origin of the four epithelial cell types. *The American Journal of Anatomy*, *141*(4), 537–561.

Condeelis, J., & Pollard, J. W. (2006). Macrophages: Obligate partners for tumor cell migration, invasion, and metastasis. *Cell*, *124*(2), 263–266.

Corcoran, R. B., Ebi, H., Turke, A. B., Coffee, E. M., Nishino, M., Cogdill, A. P., et al. (2012). EGFR-mediated re-activation of MAPK signaling contributes to insensitivity of BRAF mutant colorectal cancers to RAF inhibition with vemurafenib. *Cancer Discovery*, *2*(3), 227–235.

Coussens, L. M., Fingleton, B., & Matrisian, L. M. (2002). Matrix metalloproteinase inhibitors and cancer—Trials and tribulations. *Science*, *295*(5564), 2387–2392.

Cristobal, A., van den Toorn, H. W., van de Wetering, M., Clevers, H., Heck, A. J., & Mohammed, S. (2017). Personalized proteome profiles of healthy and tumor human colon organoids reveal both individual diversity and basic features of colorectal cancer. *Cell Reports*, *18*(1), 263–274.

de la Motte, C. A. (2011). Hyaluronan in intestinal homeostasis and inflammation: Implications for fibrosis. *American Journal of Physiology. Gastrointestinal and Liver Physiology*, *301*(6), G945–G949.

Deryugina, E. I., & Quigley, J. P. (2006). Matrix metalloproteinases and tumor metastasis. *Cancer and Metastasis Reviews*, *25*(1), 9–34.

Devy, L., & Dransfield, D. T. (2011). New strategies for the next generation of matrix-metalloproteinase inhibitors: Selectively targeting membrane-anchored MMPs with therapeutic antibodies. *Biochemistry Research International*, *2011*(7), 1–11.

Drummond, A. H., Beckett, P., Brown, P. D., Bone, E. A., Davidson, A. H., Galloway, W. A., et al. (1999). Preclinical and clinical studies of MMP inhibitors in cancer. *Annals of the New York Academy of Sciences*, *878*, 228–235.

Duerr, R. H., Taylor, K. D., Brant, S. R., Rioux, J. D., Silverberg, M. S., Daly, M. J., et al. (2006). A genome-wide association study identifies IL23R as an inflammatory bowel disease gene. *Science*, *314*(5804), 1461–1463.

Egeblad, M., Nakasone, E. S., & Werb, Z. (2010). Tumors as organs: Complex tissues that interface with the entire organism. *Developmental Cell*, *18*(6), 884–901.
Eke, I., Storch, K., Krause, M., & Cordes, N. (2013). Cetuximab attenuates its cytotoxic and radiosensitizing potential by inducing fibronectin biosynthesis. *Cancer Research*, *73*(19), 5869–5879.
Elliott, M. J., De Coppi, P., Speggiorin, S., Roebuck, D., Butler, C. R., Samuel, E., et al. (2012). Stem-cell-based, tissue engineered tracheal replacement in a child: A 2-year follow-up study. *Lancet*, *380*(9846), 994–1000.
Erez, N., Truitt, M., Olson, P., & Hanahan, D. (2010). Cancer-associated fibroblasts are activated in incipient neoplasia to orchestrate tumor-promoting inflammation in an NF-kappaB-dependent manner. *Cancer Cell*, *17*(2), 135–147.
Farin, H. F., Van Es, J. H., & Clevers, H. (2012). Redundant sources of Wnt regulate intestinal stem cells and promote formation of Paneth cells. *Gastroenterology*, *143*(6),1518–1529.e1517.
Fedarko, N. S., Jain, A., Karadag, A., Van Eman, M. R., & Fisher, L. W. (2001). Elevated serum bone sialoprotein and osteopontin in colon, breast, prostate, and lung cancer. *Clinical Cancer Research*, 7(12), 4060–4066.
Fernandez-Sanchez, M. E., Barbier, S., Whitehead, J., Bealle, G., Michel, A., Latorre-Ossa,-H., et al. (2015). Mechanical induction of the tumorigenic beta-catenin pathway by tumour growth pressure. *Nature*, *523*(7558), 92–95.
Fordham, R. P., Yui, S., Hannan, N. R., Soendergaard, C., Madgwick, A., Schweiger, P. J., et al. (2013). Transplantation of expanded fetal intestinal progenitors contributes to colon regeneration after injury. *Cell Stem Cell*, *13*(6), 734–744.
Fridlender, Z. G., Sun, J., Kim, S., Kapoor, V., Cheng, G., Ling, L., et al. (2009). Polarization of tumor-associated neutrophil phenotype by TGF-beta: "N1" versus "n2" TAN. *Cancer Cell*, *16*(3), 183–194.
Friedmann, Y., Vlodavsky, I., Aingorn, H., Aviv, A., Peretz, T., Pecker, I., et al. (2000). Expression of heparanase in normal, dysplastic, and neoplastic human colonic mucosa and stroma. Evidence for its role in colonic tumorigenesis. *The American Journal of Pathology*, *157*(4), 1167–1175.
Fukuda, M., Mizutani, T., Mochizuki, W., Matsumoto, T., Nozaki, K., Sakamaki, Y., et al. (2014). Small intestinal stem cell identity is maintained with functional Paneth cells in heterotopically grafted epithelium onto the colon. *Genes & Development*, *28*(16), 1752–1757.
Gagne, D., Groulx, J. F., Benoit, Y. D., Basora, N., Herring, E., Vachon, P. H., et al. (2010). Integrin-linked kinase regulates migration and proliferation of human intestinal cells under a fibronectin-dependent mechanism. *Journal of Cellular Physiology*, *222*(2), 387–400.
Geboes, K., Colombel, J. F., Greenstein, A., Jewell, D. P., Sandborn, W. J., Vatn, M. H., et al. (2008). Indeterminate colitis: A review of the concept—What's in a name? *Inflammatory Bowel Diseases*, *14*(6), 850–857.
Gelbmann, C. M., Mestermann, S., Gross, V., Kollinger, M., Scholmerich, J., & Falk, W. (1999). Strictures in Crohn's disease are characterised by an accumulation of mast cells colocalised with laminin but not with fibronectin or vitronectin. *Gut*, *45*(2), 210–217.
Gjorevski, N., Sachs, N., Manfrin, A., Giger, S., Bragina, M. E., Ordonez-Moran, P., et al. (2016). Designer matrices for intestinal stem cell and organoid culture. *Nature*, *539*(7630), 560–564.
Gong, Y., Scott, E., Lu, R., Xu, Y., Oh, W. K., & Yu, Q. (2013). TIMP-1 promotes accumulation of cancer associated fibroblasts and cancer progression. *PLoS One*, *8*(10), e77366.
Gregorieff, A., Pinto, D., Begthel, H., Destrée, O., Kielman, M., & Clevers, H. (2005). Expression pattern of Wnt signaling components in the adult intestine. *Gastroenterology*, *129*(2), 626–638.

Greten, F. R., Eckmann, L., Greten, T. F., Park, J. M., Li, Z.-W., Egan, L. J., et al. (2004). IKKβ links inflammation and tumorigenesis in a mouse model of colitis-associated cancer. *Cell*, *118*(3), 285–296.
Grivennikov, S. I., Greten, F. R., & Karin, M. (2010). Immunity, inflammation, and cancer. *Cell*, *140*(6), 883–899.
Groeber, F., Engelhardt, L., Lange, J., Kurdyn, S., Schmid, F. F., Rucker, C., et al. (2016). A first vascularized skin equivalent for as an alternative to animal experimentation. *ALTEX*, *33*(4), 415–422.
Guinney, J., Dienstmann, R., Wang, X., de Reyniès, A., Schlicker, A., Soneson, C., et al. (2015). The consensus molecular subtypes of colorectal cancer. *Nature Medicine*, *21*(11), 1350–1356.
Hagemann, T., Lawrence, T., McNeish, I., Charles, K. A., Kulbe, H., Thompson, R. G., et al. (2008). "Re-educating" tumor-associated macrophages by targeting NF-kappaB. *The Journal of Experimental Medicine*, *205*(6), 1261–1268.
Hanauer, S. B. (2004). Update on the etiology, pathogenesis and diagnosis of ulcerative colitis. *Nature Clinical Practice. Gastroenterology & Hepatology*, *1*(1), 26–31.
Hatoum, O. A., Heidemann, J., & Binion, D. G. (2006). The intestinal microvasculature as a therapeutic target in inflammatory bowel disease. *Annals of the New York Academy of Sciences*, *1072*, 78–97.
Hawinkels, L. J. A. C., Paauwe, M., Verspaget, H. W., Wiercinska, E., van der Zon, J. M., van der Ploeg, K., et al. (2012). Interaction with colon cancer cells hyperactivates TGF-beta signaling in cancer-associated fibroblasts. *Oncogene*, *33*(1), 97–107.
Hayashi, K., Fong, K. S., Mercier, F., Boyd, C. D., Csiszar, K., & Hayashi, M. (2004). Comparative immunocytochemical localization of lysyl oxidase (LOX) and the lysyl oxidase-like (LOXL) proteins: Changes in the expression of LOXL during development and growth of mouse tissues. *Journal of Molecular Histology*, *35*(8–9), 845–855.
He, X. C., Zhang, J., Tong, W. G., Tawfik, O., Ross, J., Scoville, D. H., et al. (2004). BMP signaling inhibits intestinal stem cell self-renewal through suppression of Wnt-beta-catenin signaling. *Nature Genetics*, *36*(10), 1117–1121.
Heldin, C. H., Rubin, K., Pietras, K., & Ostman, A. (2004). High interstitial fluid pressure—An obstacle in cancer therapy. *Nature Reviews. Cancer*, *4*(10), 806–813.
Hindryckx, P., Jairath, V., & D'Haens, G. (2016). Acute severe ulcerative colitis: From pathophysiology to clinical management. *Nature Reviews. Gastroenterology & Hepatology*, *13*(11), 654–664.
Houchen, C. W., George, R. J., Sturmoski, M. A., & Cohn, S. M. (1999). FGF-2 enhances intestinal stem cell survival and its expression is induced after radiation injury. *The American Journal of Physiology*, *276*(1 Pt. 1), G249–G258.
Houghton, A. M., Quintero, P. A., Perkins, D. L., Kobayashi, D. K., Kelley, D. G., Marconcini, L. A., et al. (2006). Elastin fragments drive disease progression in a murine model of emphysema. *The Journal of Clinical Investigation*, *116*(3), 753–759.
Humphries, A., & Wright, N. A. (2008). Colonic crypt organization and tumorigenesis. *Nature Reviews Cancer*, *8*(6), 415–424.
Hynes, R. O. (2009). The extracellular matrix: Not just pretty fibrils. *Science*, *326*(5957), 1216–1219.
Itzkowitz, S. H., & Yio, X. (2004). Inflammation and cancer IV. Colorectal cancer in inflammatory bowel disease: The role of inflammation. *American Journal of Physiology. Gastrointestinal and Liver Physiology*, *287*(1), G7–G17.
Jabaji, Z., Brinkley, G. J., Khalil, H. A., Sears, C. M., Lei, N. Y., Lewis, M., et al. (2014). Type I collagen as an extracellular matrix for the in vitro growth of human small intestinal epithelium. *PLoS One*, *9*(9), e107814.
Jin, H., Su, J., Garmy-Susini, B., Kleeman, J., & Varner, J. (2006). Integrin alpha4beta1 promotes monocyte trafficking and angiogenesis in tumors. *Cancer Research*, *66*(4), 2146–2152.

Johnson, L. A., Rodansky, E. S., Sauder, K. L., Horowitz, J. C., Mih, J. D., Tschumperlin, D. J., et al. (2013). Matrix stiffness corresponding to strictured bowel induces a fibrogenic response in human colonic fibroblasts. *Inflammatory Bowel Diseases*, *19*(5), 891–903.

Jones, P. L. (2001). Extracellular matrix and tenascin-C in pathogenesis of breast cancer. *The Lancet*, *357*(9273), 1992–1994.

Junttila, M. R., & de Sauvage, F. J. (2013). Influence of tumour micro-environment heterogeneity on therapeutic response. *Nature*, *501*(7467), 346–354.

Kalluri, R. (2016). The biology and function of fibroblasts in cancer. *Nature Publishing Group*, *16*(9), 582–598.

Kang, E., Yousefi, M., & Gruenheid, S. (2016). R-spondins are expressed by the intestinal stroma and are differentially regulated during Citrobacter rodentium- and DSS-induced colitis in mice. *PLoS One*, *11*(4), e0152859.

Karin, M. (2006). Nuclear factor-κb in cancer development and progression. *Nature*, *441*(7092), 431–436.

Kessenbrock, K., Plaks, V., & Werb, Z. (2010). Matrix metalloproteinases: Regulators of the tumor microenvironment. *Cell*, *141*(1), 52–67.

Kessler, S., Rho, H., West, G., Fiocchi, C., Drazba, J., & de la Motte, C. (2008). Hyaluronan (ha) deposition precedes and promotes leukocyte recruitment in intestinal inflammation. *Clinical and Translational Science*, *1*(1), 57–61.

Kim, H., Watkinson, J., Varadan, V., & Anastassiou, D. (2010). Multi-cancer computational analysis reveals invasion-associated variant of desmoplastic reaction involving INHBA, THBS2 and COL11A1. *BMC Medical Genomics*, *3*(1), 1420.

Kirkegaard, T., Hansen, A., Bruun, E., & Brynskov, J. (2004). Expression and localisation of matrix metalloproteinases and their natural inhibitors in fistulae of patients with Crohn's disease. *Gut*, *53*(5), 701–709.

Kirkland, S. C. (2009). Type I collagen inhibits differentiation and promotes a stem cell-like phenotype in human colorectal carcinoma cells. *British Journal of Cancer*, *101*(2), 320–326.

Koch, S., Nava, P., Addis, C., Kim, W., Denning, T. L., Li, L., et al. (2011). The Wnt antagonist Dkk1 regulates intestinal epithelial homeostasis and wound repair. *Gastroenterology*, *141*(1), 259–268. 268 e251–258.

Koizumi, M., King, N., Lobb, R., Benjamin, C., & Podolsky, D. K. (1992). Expression of vascular adhesion molecules in inflammatory bowel disease. *Gastroenterology*, *103*(3), 840–847.

Kosinski, C., Li, V. S., Chan, A. S., Zhang, J., Ho, C., Tsui, W. Y., et al. (2007). Gene expression patterns of human colon tops and basal crypts and BMP antagonists as intestinal stem cell niche factors. *Proceedings of the National Academy of Sciences of the United States of America*, *104*(39), 15418–15423.

Koutroubakis, I. E., Petinaki, E., Dimoulios, P., Vardas, E., Roussomoustakaki, M., Maniatis, A. N., et al. (2003). Serum laminin and collagen IV in inflammatory bowel disease. *Journal of Clinical Pathology*, *56*(11), 817–820.

Langlands, A. J., Almet, A. A., Appleton, P. L., Newton, I. P., Osborne, J. M., & Nathke, I. S. (2016). Paneth cell-rich regions separated by a cluster of Lgr5 + cells initiate crypt fission in the intestinal stem cell niche. *PLoS Biology*, *14*(6), e1002491.

Langner, C., Aust, D., Ensari, A., Villanacci, V., Becheanu, G., Miehlke, S., et al. (2015). Histology of microscopic colitis-review with a practical approach for pathologists. *Histopathology*, *66*(5), 613–626.

Leeb, S. N., Vogl, D., Grossmann, J., Falk, W., Scholmerich, J., Rogler, G., et al. (2004). Autocrine fibronectin-induced migration of human colonic fibroblasts. *The American Journal of Gastroenterology*, *99*(2), 335–340.

Leeb, S. N., Vogl, D., Gunckel, M., Kiessling, S., Falk, W., Goke, M., et al. (2003). Reduced migration of fibroblasts in inflammatory bowel disease: Role of inflammatory mediators and focal adhesion kinase. *Gastroenterology*, *125*(5), 1341–1354.

Lochter, A., Galosy, S., Muschler, J., Freedman, N., Werb, Z., & Bissell, M. J. (1997). Matrix metalloproteinase stromelysin-1 triggers a cascade of molecular alterations that leads to stable epithelial-to-mesenchymal conversion and a premalignant phenotype in mammary epithelial cells. *The Journal of Cell Biology*, *139*(7), 1861–1872.

Lohr, K., Sardana, H., Lee, S., Wu, F., Huso, D. L., Hamad, A. R., et al. (2012). Extracellular matrix protein lumican regulates inflammation in a mouse model of colitis. *Inflammatory Bowel Diseases*, *18*(1), 143–151.

Lotz, M. M., Nusrat, A., Madara, J. L., Ezzell, R., Wewer, U. M., & Mercurio, A. M. (1997). Intestinal epithelial restitution. Involvement of specific laminin isoforms and integrin laminin receptors in wound closure of a transformed model epithelium. *The American Journal of Pathology*, *150*(2), 747–760.

Lucero, H. A., & Kagan, H. M. (2006). Lysyl oxidase: An oxidative enzyme and effector of cell function. *Cellular and Molecular Life Sciences*, *63*(19–20), 2304–2316.

Luna, J., Masamunt, M. C., Llach, J., Delgado, S., & Sans, M. (2011). Palm oil tocotrienol rich fraction reduces extracellular matrix production by inhibiting transforming growth factor-beta1 in human intestinal fibroblasts. *Clinical Nutrition*, *30*(6), 858–864.

Madison, B. B., Braunstein, K., Kuizon, E., Portman, K., Qiao, X. T., & Gumucio, D. L. (2005). Epithelial hedgehog signals pattern the intestinal crypt-villus axis. *Development*, *132*(2), 279–289.

Magdeldin, T., Lopez-Davila, V., Pape, J., Cameron, G. W., Emberton, M., Loizidou, M., et al. (2017). Engineering a vascularised 3D in vitro model of cancer progression. *Scientific Reports*, 7, 44045.

Magro, F., Rodrigues-Pinto, E., Coelho, R., Andrade, P., Santos-Antunes, J., Lopes, S., et al. (2014). Is it possible to change phenotype progression in Crohn's disease in the era of immunomodulators? Predictive factors of phenotype progression. *The American Journal of Gastroenterology*, *109*(7), 1026–1036.

Mahida, Y. R., Beltinger, J., Makh, S., Goke, M., Gray, T., Podolsky, D. K., et al. (1997). Adult human colonic subepithelial myofibroblasts express extracellular matrix proteins and cyclooxygenase-1 and -2. *The American Journal of Physiology*, *273*(6 Pt. 1), G1341–G1348.

Mahida, Y. R., Galvin, A. M., Gray, T., Makh, S., McAlindon, M. E., Sewell, H. F., et al. (1997). Migration of human intestinal lamina propria lymphocytes, macrophages and eosinophils following the loss of surface epithelial cells. *Clinical and Experimental Immunology*, *109*(2), 377–386.

Mahoney, Z. X., Stappenbeck, T. S., & Miner, J. H. (2008). Laminin alpha 5 influences the architecture of the mouse small intestine mucosa. *Journal of Cell Science*, *121*(Pt. 15), 2493–2502.

Mak, I. W., Evaniew, N., & Ghert, M. (2014). Lost in translation: Animal models and clinical trials in cancer treatment. *American Journal of Translational Research*, *6*, 114–118.

Malik, R., Lelkes, P. I., & Cukierman, E. (2015). Biomechanical and biochemical remodeling of stromal extracellular matrix in cancer. *Trends in Biotechnology*, *33*(4), 230–236.

Mantovani, A., Romero, P., Palucka, A. K., & Marincola, F. M. (2008). Tumour immunity: Effector response to tumour and role of the microenvironment. *Lancet*, *371*(9614), 771–783.

Medema, J. P., & Vermeulen, L. (2011). Microenvironmental regulation of stem cells in intestinal homeostasis and cancer. *Nature*, *474*(7351), 318–326.

Metzger, M., Caldwell, C., Barlow, A. J., Burns, A. J., & Thapar, N. (2009). Enteric nervous system stem cells derived from human gut mucosa for the treatment of aganglionic gut disorders. *Gastroenterology*, *136*(7). 2214–2225 e2211–2213.

Middleton, S. J., & Jamieson, N. V. (2005). The current status of small bowel transplantation in the UK and internationally. *Gut*, *54*(11), 1650–1657.

Miyoshi, H., Ajima, R., Luo, C. T., Yamaguchi, T. P., & Stappenbeck, T. S. (2012). Wnt5a potentiates TGF-beta signaling to promote colonic crypt regeneration after tissue injury. *Science*, *338*(6103), 108–113.

Moll, C., Reboredo, J., Schwarz, T., Appelt, A., Schürlein, S., Walles, H., et al. (2013). Tissue engineering of a human 3D in vitro tumor test system. *Journal of Visualized Experiments*, *78*, e50460.

Monboisse, J. C., Oudart, J. B., Brezillon, S., Brassart, B., Ramont, L., Maquart, F. X., et al. (2013). Control of tumor progression by extracellular matrix molecule fragments, the matrikines. *Journal of Carcinogenesis & Mutagenesis*, *4*(3), 1–8.

Monteleone, I., Federici, M., Sarra, M., Franze, E., Casagrande, V., Zorzi, F., et al. (2012). Tissue inhibitor of metalloproteinase-3 regulates inflammation in human and mouse intestine. *Gastroenterology*, *143*(5), 1277–1287.e1271–1274.

Murphy, A. G., Casey, R., Maguire, A., Tosetto, M., Butler, C. T., Conroy, E., et al. (2016). Preclinical validation of the small molecule drug quininib as a novel therapeutic for colorectal cancer. *Nature Publishing Group*, *6*(1), 1374.

Naba, A., Clauser, K. R., Whittaker, C. A., Carr, S. A., Tanabe, K. K., & Hynes, R. O. (2014). Extracellular matrix signatures of human primary metastatic colon cancers and their metastases to liver. *BMC Cancer*, *14*(1), 518. 512.

Nagy, N., Bronckart, Y., Camby, I., Legendre, H., Lahm, H., Kaltner, H., et al. (2002). Galectin-8 expression decreases in cancer compared with normal and dysplastic human colon tissue and acts significantly on human colon cancer cell migration as a suppressor. *Gut*, *50*(3), 392–401.

Nakamura, M., Inufusa, H., Adachi, T., Aga, M., Kurimoto, M., Nakatani, Y., et al. (1999). Involvement of galectin-3 expression in colorectal cancer progression and metastasis. *International Journal of Oncology*, *15*(1), 143–148.

Nava, P., Koch, S., Laukoetter, M. G., Lee, W. Y., Kolegraff, K., Capaldo, C. T., et al. (2010). Interferon-gamma regulates intestinal epithelial homeostasis through converging beta-catenin signaling pathways. *Immunity*, *32*(3), 392–402.

Nietzer, S., Baur, F., Sieber, S., Hansmann, J., Schwarz, T., Stoffer, C., et al. (2016). Mimicking metastases including tumor stroma: A new technique to generate a three-dimensional colorectal cancer model based on a biological decellularized intestinal scaffold. *Tissue Engineering. Part C, Methods*, *22*(7), 621–635.

Nyga, A., Cheema, U., & Loizidou, M. (2011). 3D tumour models: Novel in vitro approaches to cancer studies. *Journal of Cell Communication and Signaling*, *5*(3), 239–248.

Nyga, A., Loizidou, M., Emberton, M., & Cheema, U. (2013). A novel tissue engineered three-dimensional in vitro colorectal cancer model. *Acta Biomaterialia*, *9*(8), 7917–7926.

Ogura, Y., Bonen, D. K., Inohara, N., Nicolae, D. L., Chen, F. F., Ramos, R., et al. (2001). A frameshift mutation in NOD2 associated with susceptibility to Crohn's disease. *Nature*, *411*(6837), 603–606.

Ohtani, H., Motohashi, H., Sato, H., Seiki, M., & Nagura, H. (1996). Dual over-expression pattern of membrane-type metalloproteinase-1 in cancer and stromal cells in human gastrointestinal carcinoma revealed by in situ hybridization and immunoelectron microscopy. *International Journal of Cancer*, *68*(5), 565–570.

Oku, Y., Shimoji, T., Takifuji, K., Hotta, T., Yokoyama, S., Matsuda, K., et al. (2008). Identification of the molecular mechanisms for dedifferentiation at the invasion front of colorectal cancer by a gene expression analysis. *Clinical Cancer Research: An Official Journal of the American Association for Cancer Research*, *14*(22), 7215–7222.

Okuno, T., Andoh, A., Bamba, S., Araki, Y., Fujiyama, Y., Fujiyama, M., et al. (2002). Interleukin-1beta and tumor necrosis factor-alpha induce chemokine and matrix metalloproteinase gene expression in human colonic subepithelial myofibroblasts. *Scandinavian Journal of Gastroenterology*, *37*(3), 317–324.

Ootani, A., Li, X., Sangiorgi, E., Ho, Q. T., Ueno, H., Toda, S., et al. (2009). Sustained in vitro intestinal epithelial culture within a Wnt-dependent stem cell niche. *Nature Medicine, 15*(6), 701–706.

O'Sullivan, S., Gilmer, J. F., & Medina, C. (2015). Matrix metalloproteinases in inflammatory bowel disease: An update. *Mediators of Inflammation, 2015*, 964131.

Ott, H. C., Clippinger, B., Conrad, C., Schuetz, C., Pomerantseva, I., Ikonomou, L., et al. (2010). Regeneration and orthotopic transplantation of a bioartificial lung. *Nature Medicine, 16*(8), 927–933.

Ott, H. C., Matthiesen, T. S., Goh, S. K., Black, L. D., Kren, S. M., Netoff, T. I., et al. (2008). Perfusion-decellularized matrix: Using nature's platform to engineer a bioartificial heart. *Nature Medicine, 14*(2), 213–221.

Perel, P., Roberts, I., Sena, E., Wheble, P., Briscoe, C., Sandercock, P., et al. (2007). Comparison of treatment effects between animal experiments and clinical trials: Systematic review. *BMJ (Clinical Research ed.), 334*(7586), 197.

Petersen, T. H., Calle, E. A., Zhao, L., Lee, E. J., Gui, L., Raredon, M. B., et al. (2010). Tissue-engineered lungs for in vivo implantation. *Science, 329*(5991), 538–541.

Petrey, A. C., & de la Motte, C. A. (2017). The extracellular matrix in IBD: A dynamic mediator of inflammation. *Current Opinion in Gastroenterology, 33*(4), 234–238.

Pickup, M. W., Mouw, J. K., & Weaver, V. M. (2014). The extracellular matrix modulates the hallmarks of cancer. *EMBO Reports, 15*(12), 1243–1253.

Pocza, P., Suli-Vargha, H., Darvas, Z., & Falus, A. (2008). Locally generated VGVAPG and VAPG elastin-derived peptides amplify melanoma invasion via the galectin-3 receptor. *International Journal of Cancer, 122*(9), 1972–1980.

Podolsky, D. K. (2002). Inflammatory bowel disease. *The New England Journal of Medicine, 347*(6), 417–429.

Poghosyan, T., Sfeir, R., Michaud, L., Bruneval, P., Domet, T., Vanneaux, V., et al. (2015). Circumferential esophageal replacement using a tube-shaped tissue-engineered substitute: An experimental study in minipigs. *Surgery, 158*(1), 266–277.

Pollard, J. W. (2009). Trophic macrophages in development and disease. *Nature Reviews. Immunology, 9*(4), 259–270.

Porte, H., Chastre, E., Prevot, S., Nordlinger, B., Empereur, S., Basset, P., et al. (1995). Neoplastic progression of human colorectal cancer is associated with overexpression of the stromelysin-3 and BM-40/SPARC genes. *International Journal of Cancer, 64*(1), 70–75.

Potten, C. S., Kovacs, L., & Hamilton, E. (1974). Continuous labelling studies on mouse skin and intestine. *Cell and Tissue Kinetics*, 7(3), 271–283.

Potten, C. S., Owen, G., Hewitt, D., Chadwick, C. A., Hendry, H., Lord, B. I., et al. (1995). Stimulation and inhibition of proliferation in the small intestinal crypts of the mouse after in vivo administration of growth factors. *Gut, 36*(6), 864–873.

Prahallad, A., Sun, C., Huang, S., Di Nicolantonio, F., Salazar, R., Zecchin, D., et al. (2012). Unresponsiveness of colon cancer to BRAF(V600E) inhibition through feedback activation of EGFR. *Nature, 483*(7387), 100–103.

Provenzano, P. P., Eliceiri, K. W., Campbell, J. M., Inman, D. R., White, J. G., & Keely, P. J. (2006). Collagen reorganization at the tumor-stromal interface facilitates local invasion. *BMC Medicine, 4*(1), 38.

Psaila, B., & Lyden, D. (2009). The metastatic niche: Adapting the foreign soil. *Nature Reviews. Cancer, 9*(4), 285–293.

Pucilowska, J. B., Williams, K. L., & Lund, P. K. (2000). Fibrogenesis. IV. Fibrosis and inflammatory bowel disease: Cellular mediators and animal models. *American Journal of Physiology. Gastrointestinal and Liver Physiology, 279*(4), G653–G659.

Pull, S. L., Doherty, J. M., Mills, J. C., Gordon, J. I., & Stappenbeck, T. S. (2005). Activated macrophages are an adaptive element of the colonic epithelial progenitor niche necessary for regenerative responses to injury. *Proceedings of the National Academy of Sciences of the United States of America, 102*(1), 99–104.

Qi, Z., Li, Y., Zhao, B., Xu, C., Liu, Y., Li, H., et al. (2016). BMP restricts stemness of intestinal Lgr5 + stem cells by directly suppressing their signature genes. *Nature Communications*, 7, 1–14.

Quaroni, A., Isselbacher, K. J., & Ruoslahti, E. (1978). Fibronectin synthesis by epithelial crypt cells of rat small intestine. *Proceedings of the National Academy of Sciences of the United States of America*, *75*(11), 5548–5552.

Reya, T., & Clevers, H. (2005). Wnt signalling in stem cells and cancer. *Nature*, *434*, 843–850.

Rieder, F., Brenmoehl, J., Leeb, S., Scholmerich, J., & Rogler, G. (2007). Wound healing and fibrosis in intestinal disease. *Gut*, *56*(1), 130–139.

Rieder, F., & Fiocchi, C. (2009). Intestinal fibrosis in IBD—A dynamic, multifactorial process. *Nature Reviews. Gastroenterology & Hepatology*, *6*(4), 228–235.

Riehl, T. E., Foster, L., & Stenson, W. F. (2012). Hyaluronic acid is radioprotective in the intestine through a TLR4 and COX-2-mediated mechanism. *American Journal of Physiology. Gastrointestinal and Liver Physiology*, *302*(3), G309–G316.

Rijcken, E., Mennigen, R. B., Schaefer, S. D., Laukoetter, M. G., Anthoni, C., Spiegel, H. U., et al. (2007). PECAM-1 (CD 31) mediates transendothelial leukocyte migration in experimental colitis. *American Journal of Physiology. Gastrointestinal and Liver Physiology*, *293*(2), G446–G452.

Sailaja, B. S., He, X. C., & Li, L. (2016). The regulatory niche of intestinal stem cells. *The Journal of Physiology*, *594*(17), 4827–4836.

Sangiorgi, E., & Capecchi, M. R. (2008). Bmi1 is expressed in vivo in intestinal stem cells. *Nature Genetics*, *40*(7), 915–920.

Sasaki, N., Sachs, N., Wiebrands, K., Ellenbroek, S. I., Fumagalli, A., Lyubimova, A., et al. (2016). Reg4 + deep crypt secretory cells function as epithelial niche for Lgr5 + stem cells in colon. *Proceedings of the National Academy of Sciences of the United States of America*, *113*(37), E5399–E5407.

Sato, T., Stange, D. E., Ferrante, M., Vries, R. G., Van Es, J. H., Van den Brink, S., et al. (2011). Long-term expansion of epithelial organoids from human colon, adenoma, adenocarcinoma, and Barrett's epithelium. *Gastroenterology*, *141*(5), 1762–1772.

Sato, T., van Es, J. H., Snippert, H. J., Stange, D. E., Vries, R. G., van den Born, M., et al. (2011). Paneth cells constitute the niche for Lgr5 stem cells in intestinal crypts. *Nature*, *469*(7330), 415–418.

Sato, T., Vries, R. G., Snippert, H. J., van de Wetering, M., Barker, N., Stange, D. E., et al. (2009). Single Lgr5 stem cells build crypt-villus structures in vitro without a mesenchymal niche. *Nature*, *459*(7244), 262–265.

Scharl, M., & Rogler, G. (2014). Pathophysiology of fistula formation in Crohn's disease. *World Journal of Gastrointestinal Pathophysiology*, *5*(3), 205–212.

Scheibner, K. A., Lutz, M. A., Boodoo, S., Fenton, M. J., Powell, J. D., & Horton, M. R. (2006). Hyaluronan fragments act as an endogenous danger signal by engaging TLR2. *Journal of Immunology*, *177*(2), 1272–1281.

Scherer, R. L., McIntyre, J. O., & Matrisian, L. M. (2008). Imaging matrix metalloproteinases in cancer. *Cancer and Metastasis Reviews*, *27*(4), 679–690.

Schmehl, K., Florian, S., Jacobasch, G., Salomon, A., & Korber, J. (2000). Deficiency of epithelial basement membrane laminin in ulcerative colitis affected human colonic mucosa. *International Journal of Colorectal Disease*, *15*(1), 39–48.

Schuijers, J., & Clevers, H. (2012). Adult mammalian stem cells: The role of Wnt, Lgr5 and R-spondins. *The EMBO Journal*, *31*(12), 2685–2696.

Schweinlin, M., Wilhelm, S., Schwedhelm, I., Hansmann, J., Rietscher, R., Jurowich, C., et al. (2016). Development of an advanced primary human in vitro model of the small intestine. *Tissue Engineering. Part C, Methods*, *22*(9), 873–883.

Seliktar, D., Nerem, R. M., & Galis, Z. S. (2001). The role of matrix metalloproteinase-2 in the remodeling of cell-seeded vascular constructs subjected to cyclic strain. *Annals of Biomedical Engineering*, *29*(11), 923–934.

Seno, H., Miyoshi, H., Brown, S. L., Geske, M. J., Colonna, M., & Stappenbeck, T. S. (2009). Efficient colonic mucosal wound repair requires Trem2 signaling. *Proceedings of the National Academy of Sciences of the United States of America*, *106*(1), 256–261.
Shi, L., Ramsay, S., Ermis, R., & Carson, D. (2012). In vitro and in vivo studies on matrix metalloproteinases interacting with small intestine submucosa wound matrix. *International Wound Journal*, *9*(1), 44–53.
Shimshoni, E., Yablecovitch, D., Baram, L., Dotan, I., & Sagi, I. (2015). ECM remodelling in IBD: Innocent bystander or partner in crime? The emerging role of extracellular molecular events in sustaining intestinal inflammation. *Gut*, *64*(3), 367–372.
Simo, P., Simon-Assmann, P., Bouziges, F., Leberquier, C., Kedinger, M., Ekblom, P., et al. (1991). Changes in the expression of laminin during intestinal development. *Development*, *112*(2), 477–487.
Simon-Assmann, P., Lefebvre, O., Bellissent-Waydelich, A., Olsen, J., Orian-Rousseau, V., & De Arcangelis, A. (1998). The laminins: Role in intestinal morphogenesis and differentiation. *Annals of the New York Academy of Sciences*, *859*, 46–64.
Simpson, R. M., Meran, S., Thomas, D., Stephens, P., Bowen, T., Steadman, R., et al. (2009). Age-related changes in pericellular hyaluronan organization leads to impaired dermal fibroblast to myofibroblast differentiation. *The American Journal of Pathology*, *175*(5), 1915–1928.
Sneddon, J. B., Zhen, H. H., Montgomery, K., van de Rijn, M., Tward, A. D., West, R., et al. (2006). Bone morphogenetic protein antagonist gremlin 1 is widely expressed by cancer-associated stromal cells and can promote tumor cell proliferation. *Proceedings of the National Academy of Sciences of the United States of America*, *103*(40), 14842–14847.
Sobrino, A., Phan, D. T., Datta, R., Wang, X., Hachey, S. J., Romero-Lopez, M., et al. (2016). 3D microtumors in vitro supported by perfused vascular networks. *Scientific Reports*, *6*, 31589.
Solli, A. I., Fadnes, B., Winberg, J. O., Uhlin-Hansen, L., & Hadler-Olsen, E. (2013). Tissue- and cell-specific co-localization of intracellular gelatinolytic activity and matrix metalloproteinase 2. *The Journal of Histochemistry and Cytochemistry*, *61*(6), 444–461.
Song, X., Dai, D., He, X., Zhu, S., Yao, Y., Gao, H., et al. (2015). Growth factor FGF2 cooperates with interleukin-17 to repair intestinal epithelial damage. *Immunity*, *43*(3), 488–501.
Song, J. J., Guyette, J. P., Gilpin, S. E., Gonzalez, G., Vacanti, J. P., & Ott, H. C. (2013). Regeneration and experimental orthotopic transplantation of a bioengineered kidney. *Nature Medicine*, *19*(5), 646–651.
Soroosh, A., Albeiroti, S., West, G. A., Willard, B., Fiocchi, C., & de la Motte, C. A. (2016). Crohn's disease fibroblasts overproduce the novel protein KIAA1199 to create proinflammatory hyaluronan fragments. *Cellular and Molecular Gastroenterology and Hepatology*, *2*(3), 358–368.e354.
Spaderna, S., Schmalhofer, O., Hlubek, F., Berx, G., Eger, A., Merkel, S., et al. (2006). A transient, EMT-linked loss of basement membranes indicates metastasis and poor survival in colorectal cancer. *Gastroenterology*, *131*(3), 830–840.
Sparmann, A., & Bar-Sagi, D. (2004). Ras-induced interleukin-8 expression plays a critical role in tumor growth and angiogenesis. *Cancer Cell*, *6*(5), 447–458.
Stahle-Backdahl, M., Maim, J., Veress, B., Benoni, C., Bruce, K., & Egesten, A. (2000). Increased presence of eosinophilic granulocytes expressing transforming growth factor-beta1 in collagenous colitis. *Scandinavian Journal of Gastroenterology*, *35*(7), 742–746.
Stallmach, A., Schuppan, D., Riese, H. H., Matthes, H., & Riecken, E. O. (1992). Increased collagen type III synthesis by fibroblasts isolated from strictures of patients with Crohn's disease. *Gastroenterology*, *102*(6), 1920–1929.

Strup-Perrot, C., Vozenin-Brotons, M. C., Vandamme, M., Benderitter, M., & Mathe, D. (2006). Expression and activation of MMP-2, -3, -9, -14 are induced in rat colon after abdominal x-irradiation. *Scandinavian Journal of Gastroenterology*, *41*(1), 60–70.

Stzepourginski, I., Nigro, G., Jacob, J. M., Dulauroy, S., Sansonetti, P. J., Eberl, G., et al. (2017). CD34+ mesenchymal cells are a major component of the intestinal stem cells niche at homeostasis and after injury. *Proceedings of the National Academy of Sciences of the United States of America*, *114*(4), E506–E513.

Sun, C., Wang, L., Huang, S., Heynen, G. J., Prahallad, A., Robert, C., et al. (2014). Reversible and adaptive resistance to BRAF(V600E) inhibition in melanoma. *Nature*, *508*(7494), 118–122.

Swindle, C. S., Tran, K. T., Johnson, T. D., Banerjee, P., Mayes, A. M., Griffith, L., et al. (2001). Epidermal growth factor (EGF)-like repeats of human tenascin-C as ligands for EGF receptor. *The Journal of Cell Biology*, *154*(2), 459–468.

Taha, Y., Raab, Y., Larsson, A., Carlson, M., Loof, L., Gerdin, B., et al. (2003). Mucosal secretion and expression of basic fibroblast growth factor in patients with collagenous colitis. *The American Journal of Gastroenterology*, *98*(9), 2011–2017.

Takaku, K., Oshima, M., Miyoshi, H., Matsui, M., Seldin, M. F., & Taketo, M. M. (1998). Intestinal tumorigenesis in compound mutant mice of both Dpc4 (Smad4) and Apc genes. *Cell*, *92*(5), 645–656.

Taniwaki, K., Fukamachi, H., Komori, K., Ohtake, Y., Nonaka, T., Sakamoto, T., et al. (2007). Stroma-derived matrix metalloproteinase (MMP)-2 promotes membrane type 1-MMP-dependent tumor growth in mice. *Cancer Research*, *67*(9), 4311–4319.

Tian, H., Biehs, B., Warming, S., Leong, K. G., Rangell, L., Klein, O. D., et al. (2012). A reserve stem cell population in small intestine renders Lgr5-positive cells dispensable. *Nature*, *478*(7368), 255–259.

Tilghman, R. W., & Parsons, J. T. (2008). Focal adhesion kinase as a regulator of cell tension in the progression of cancer. *Seminars in Cancer Biology*, *18*(1), 45–52.

Torre, L. A., Bray, F., Siegel, R. L., Ferlay, J., Lortet-Tieulent, J., & Jemal, A. (2015). Global cancer statistics, 2012. *CA: A Cancer Journal for Clinicians*, *65*(2), 87–108.

Totonelli, G., Maghsoudlou, P., Garriboli, M., Riegler, J., Orlando, G., Burns, A. J., et al. (2012). A rat decellularized small bowel scaffold that preserves villus-crypt architecture for intestinal regeneration. *Biomaterials*, *33*(12), 3401–3410.

Vaday, G. G., Hershkoviz, R., Rahat, M. A., Lahat, N., Cahalon, L., & Lider, O. (2000). Fibronectin-bound TNF-alpha stimulates monocyte matrix metalloproteinase-9 expression and regulates chemotaxis. *Journal of Leukocyte Biology*, *68*(5), 737–747.

van de Wetering, M., Francies, H. E., Francis, J. M., Bounova, G., Iorio, F., Pronk, A., et al. (2015). Prospective derivation of a living organoid biobank of colorectal cancer patients. *Cell*, *161*(4), 933–945.

van Es, J. H., van Gijn, M. E., Riccio, O., van den Born, M., Vooijs, M., Begthel, H., et al. (2005). Notch/gamma-secretase inhibition turns proliferative cells in intestinal crypts and adenomas into goblet cells. *Nature*, *435*(7044), 959–963.

Van Landeghem, L., Santoro, M. A., Mah, A. T., Krebs, A. E., Dehmer, J. J., McNaughton, K. K., et al. (2015). IGF1 stimulates crypt expansion via differential activation of 2 intestinal stem cell populations. *The FASEB Journal*, *29*(7), 2828–2842.

Vanderhoof, J. A., & Langnas, A. N. (1997). Short-bowel syndrome in children and adults. *Gastroenterology*, *113*(5), 1767–1778.

Vermeulen, L., De Sousa E Melo, F., van der Heijden, M., Cameron, K., de Jong, J. H., Borovski, T., et al. (2010). Wnt activity defines colon cancer stem cells and is regulated by the microenvironment. *Nature Publishing Group*, *12*(5), 468–476.

Verrecchia, F., Chu, M. L., & Mauviel, A. (2001). Identification of novel TGF-beta/Smad gene targets in dermal fibroblasts using a combined cDNA microarray/promoter transactivation approach. *The Journal of Biological Chemistry*, *276*(20), 17058–17062.

Voloshanenko, O., Erdmann, G., Dubash, T. D., Augustin, I., Metzig, M., Moffa, G., et al. (2013). Wnt secretion is required to maintain high levels of Wnt activity in colon cancer cells. *Nature Communications*, *4*, 1–13.

Wang, Y., Gunasekara, D. B., Reed, M. I., DiSalvo, M., Bultman, S. J., Sims, C. E., et al. (2017). A microengineered collagen scaffold for generating a polarized crypt-villus architecture of human small intestinal epithelium. *Biomaterials*, *128*, 44–55.

Wang, L., Murthy, S. K., Barabino, G. A., & Carrier, R. L. (2010). Synergic effects of crypt-like topography and ECM proteins on intestinal cell behavior in collagen based membranes. *Biomaterials*, *31*(29), 7586–7598.

Wang, Y. D., & Yan, P. Y. (2006). Expression of matrix metalloproteinase-1 and tissue inhibitor of metalloproteinase-1 in ulcerative colitis. *World Journal of Gastroenterology*, *12*(37), 6050–6053.

Weathington, N. M., van Houwelingen, A. H., Noerager, B. D., Jackson, P. L., Kraneveld, A. D., Galin, F. S., et al. (2006). A novel peptide CXCR ligand derived from extracellular matrix degradation during airway inflammation. *Nature Medicine*, *12*(3), 317–323.

Webber, J., Jenkins, R. H., Meran, S., Phillips, A., & Steadman, R. (2009). Modulation of TGFbeta1-dependent myofibroblast differentiation by hyaluronan. *The American Journal of Pathology*, *175*(1), 148–160.

Weber, C. E., Kothari, A. N., Wai, P. Y., Li, N. Y., Driver, J., Zapf, M. A., et al. (2015). Osteopontin mediates an MZF1-TGF-beta1-dependent transformation of mesenchymal stem cells into cancer-associated fibroblasts in breast cancer. *Oncogene*, *34*(37), 4821–4833.

Wolf, K., te Lindert, M., Krause, M., Alexander, S., te Riet, J., Willis, A. L., et al. (2013). Physical limits of cell migration: Control by ECM space and nuclear deformation and tuning by proteolysis and traction force. *The Journal of Cell Biology*, *201*(7), 1069–1084.

Wolpert, S., Wong, M. L., & Bass, B. L. (1996). Matrix alters the proliferative response of enterocytes to growth factors. *American Journal of Surgery*, *171*(1), 109–112.

Wolpert, S. I., Wong, M. D., Wang, J. Y., & Bass, B. L. (1996). Epithelial-matrix interactions: Laminin downregulates enterocyte EGF receptor and IGF-I receptor expression. *The Journal of Surgical Research*, *63*(1), 345–348.

Wood, L. D., Parsons, D. W., Jones, S., Lin, J., Sjöblom, T., Leary, R. J., et al. (2007). The genomic landscapes of human breast and colorectal cancers. *Science*, *318*(5853), 1108–1113.

Wynn, T. A., & Ramalingam, T. R. (2012). Mechanisms of fibrosis: Therapeutic translation for fibrotic disease. *Nature Medicine*, *18*(7), 1028–1040.

Xavier, R. J., & Podolsky, D. K. (2007). Unravelling the pathogenesis of inflammatory bowel disease. *Nature*, *448*(7152), 427–434.

Yan, K. S., Janda, C. Y., Chang, J., Zheng, G. X. Y., Larkin, K. A., Luca, V. C., et al. (2017). Non-equivalence of Wnt and R-spondin ligands during Lgr5+ intestinal stem-cell self-renewal. *Nature Publishing Group*, *545*(7653), 238–242.

Yue, B. (2014). Biology of the extracellular matrix: An overview. *Journal of Glaucoma*, *23*(8 Suppl. 1), S20–S23.

Yui, S., Nakamura, T., Sato, T., Nemoto, Y., Mizutani, T., Zheng, X., et al. (2012). Functional engraftment of colon epithelium expanded in vitro from a single adult Lgr5(+) stem cell. *Nature Medicine*, *18*(4), 618–623.

Kolibri 360 GmbH
Ferdinand-Jühlke-Straße 7
99095 Erfurt, Deutschland
produktsicherheit@kolibri360.de